普通高等教育"十三五"规划教材

材料工程导论（双语）

主　编　傅小明　蒋　萍
副主编　杨在志　李金涛　孙　虎

南京大学出版社

图书在版编目(CIP)数据

材料工程导论：双语 / 傅小明，蒋萍主编. —南京：南京大学出版社，2017.11
 ISBN 978-7-305-19575-4

Ⅰ. ①材… Ⅱ. ①傅… ②蒋… Ⅲ. ①工程材料 Ⅳ. ①TB3

中国版本图书馆 CIP 数据核字(2017)第 274166 号

出版发行　南京大学出版社
社　　址　南京市汉口路 22 号　　邮　编　210093
出 版 人　金鑫荣

书　　名　材料工程导论（双语）
主　　编　傅小明　蒋　萍
副 主 编　杨在志　李金涛　孙　虎
责任编辑　方巧真　王日俊

照　　排　南京南琳图文制作有限公司
印　　刷　南京大众新科技印刷有限公司
开　　本　787×1092　1/16　印张 20　字数 438 千
版　　次　2017 年 11 月第 1 版　2017 年 11 月第 1 次印刷
ISBN　978-7-305-19575-4
定　　价　56.00 元

网址：http://www.njupco.com
官方微博：http://weibo.com/njupco
官方微信号：njupress
销售咨询热线：(025) 83594756

* 版权所有，侵权必究
* 凡购买南大版图书，如有印装质量问题，请与所购
　图书销售部门联系调换

前 言

随着人类文明的进步和科学技术的发展,材料已成为国民经济的三大支柱产业之一。为了适应材料科学与技术的发展,培养学生及时跟踪国际材料科学与技术发展前沿的能力,使学生成为材料科学领域的创新型复合人才,因而国内许多高校面向材料类本科生开设了双语专业课程。

目前,材料类双语专业课程《材料概论(双语)》或《材料科学与工程导论(双语)》教材尽管也有出版,但存在如下的主要问题:(1) 系统性不强,即知识点不全,不利于学生系统和全面的掌握相关知识;(2) 部分教材有少许课后练习题,但是没有参考答案,这样不利于学生课后对学习效果的检测和评价;(3) 更有甚者是半中文半英文,这样更不利于学生学习利用纯正的英语表述专业知识。这样难以满足培养应用型人才教学的要求。

编者基于上述三个问题得到如下基本认识:一是现代材料类高素质创新型人才标准——国际化、工程化和复合化的要求;二是基础应用性材料科学知识体系:科学性、系统性和简易性的特色;三是各类不同层次高校学生的教学要求。本教材针对低年级材料类大学生的实际条件和需求,在满足国际化专业人才语言交流能力的基本前提下,特别强调了专业基础知识体系的科学化、系统化和简易化,从而新编了《材料科学导论(双语)》和《材料工程导论(双语)》系列教材,以满足新时代各类特别是技术应用型材料科学与工程学科专业教学的需要。

本套教材具有如下特点:

1. 将材料学知识英语化

用英语语言思维构筑学生的国际化视野和专业语言交流能力。

2. 对知识性体系简易化

在保持教学内容的科学性、系统性的前提下,做到学生理解和掌握的简易性(通俗性),即突出"真、全、简"三个字。

3. 优化教学流程和效果

(1) 任务驱动教学

每章主要包括主要知识点的介绍、专业词汇的注解和课后练习题(附参考答案),充分体现了教学内容的实用性,有助于提高学生牢固掌握本章知识点的实践能力。

(2) 教材定位准确

本套教材针对材料类学生学习专业基础课后开设的课程,有助于学生应用英语去描述自己本专业的材料学知识。

(3) 内容结构合理

本套教材内容由浅入深，循序渐进，符合读者认识事物的规律性。同时，也便于教学的组织、实施和考核，有利于教学效果的巩固和教学质量的提高。

《材料工程导论(双语)》教材是由宿迁学院材料工程系傅小明副教授和江苏大学外国语学院蒋萍老师担任主编。傅小明老师编写绪论(第1章)、材料工程基础篇(第2章流体力学基础、第3章传热学基础和第4章扩散理论)、材料的设计与模拟篇(第5章材料的计算机辅助设计与制造)、材料的制备与加工篇(第6章粉末冶金、第10章陶瓷、第15章纳米复合材料)和材料的测试与分析篇(第16章材料的分析技术)，以及全书的统稿；杨在志老师编写材料的制备与加工篇(第7章铸造与成型、第8章焊接、第9章热处理和第11章聚合物材料)；李金涛老师编写材料的制备与加工篇(第12章金属基复合材料、第13章陶瓷基复合材料、第14章聚合物基复合材料)；孙虎老师编写材料的应用篇(第17章材料的应用)。蒋萍老师编写全书专业术语和专业词汇的注解，以及全书习题及其参考答案。

本书在编写过程中得到了兰州理工大学博士生导师马勤教授悉心的指导，在此特表感谢。

由于编者水平有限，经验不足，书中难免有不足之处，恳请专家、学者和广大读者批评指正。

编　者

2017 年 09 月

Contents

Chapter 1 Introduction ... 1
1.1 Historical Perspective ... 1
1.2 Definition of Materials ... 4
1.3 Definition of Engineering ... 4
1.4 Materials and Engineering ... 5
1.5 Selection of Materials ... 9
    1.5.1 Service Requirements ... 9
    1.5.2 Fabrication Requirements ... 9
    1.5.3 Economic Requirements ... 10
    1.5.4 Materials That Are Available to the Engineer ... 10

Part I Foundation of Material Engineering

Chapter 2 Foundation of Fluid Mechanics ... 17
2.1 Characteristics of Flow ... 17
    2.1.1 A Method to Describe Fluid Motion ... 17
    2.1.2 Laminar and Turbulent Flows ... 18
    2.1.3 Steady and Unsteady Flows ... 18
    2.1.4 Temporal Mean ... 20
    2.1.5 Uniform and Nonuniform Flows ... 20
    2.1.6 Rotational and Irrotational Flows ... 20
2.2 Concepts of Flow ... 21
    2.2.1 Flow Dimensionality ... 21
    2.2.2 Path Line and Streamline ... 21
    2.2.3 Stream Tube, Ministream Tube and Total Flow ... 22
    2.2.4 Cross Section, Flow Rate and Average Velocity ... 23
2.3 Reynolds Transport Equation ... 25
    2.3.1 System ... 25
    2.3.2 Control Volume ... 25
    2.3.3 Reynolds Transport Equation ... 26

Contents

    2.3.4 Application of the Reynolds Transport Equation 28
2.4 Continuity Equation 30
    2.4.1 Steady Flow Continuity Equation of One-Dimensional Ministream Tube 30
    2.4.2 Total Flow Continuity Equation for One-Dimensional Steady Flow ... 31
    2.4.3 Three-Dimensional Differential Continuity Equation 32
2.5 Bernoulli Equation 33
    2.5.1 Euler's Equation in One-Dimensional Flow for Ideal Fluid 33
    2.5.2 Bernoulli's Equation 34
    2.5.3 Geometry Expression of Bernoulli's Equation 36
2.6 Flow Losses and Steady Flow Energy Equation 37
    2.6.1 Reversible and Irreversible Processes 37
    2.6.2 Energy Equation in Differential Form for a Steady Flow 38
    2.6.3 Energy Equation for a Reversible Process 38
    2.6.4 Energy Equation for an Irreversible Process 39
    2.6.5 Bernoulli's Equation for Real System 39
2.7 Application of the Energy Equation to Steady Fluid-Flow Situations ... 40
    2.7.1 Gradually Varied Flow 40
    2.7.2 Bernoulli's Equation for the Real-Fluid Total Flow 42
    2.7.3 Applications of Bernoulli's Equation 43
2.8 Applications of the Linear-Momentum Equation 44

Chapter 3 Foundation of Heat Transfer 49
3.1 Introduction 49
3.2 Conduction Heat Transfer 50
3.3 Thermal Conductivity 54
3.4 Convection Heat Transfer 59
3.5 Radiation Heat Transfer 62

Chapter 4 Diffusion Theory 66
4.1 Introduction 66
4.2 Diffusion Mechanisms 68
    4.2.1 Vacancy Diffusion 68
    4.2.2 Interstitial Diffusion 69
4.3 Steady-State Diffusion 70
4.4 Nonsteady-State Diffusion 71
4.5 Factors That Influence Diffusion 74
    4.5.1 Diffusing Species 74
    4.5.2 Temperature 74

Part II Design and Simulation of Materials

Chapter 5 Computer-Aided Design and Manufacturing of Materials ... 79

5.1 Introduction ... 79
- 5.1.1 Material Forming and Manufacturing ... 79
- 5.1.2 Design Process and Concurrent Engineering ... 81
- 5.1.3 Design for Manufacture and Assembly ... 85
- 5.1.4 Selecting Materials ... 87
- 5.1.5 Selecting Manufacturing Processes ... 87
- 5.1.6 Computer Integrated Manufacturing ... 89
- 5.1.7 Machine Control Systems ... 89
- 5.1.8 Computer Technology ... 90
- 5.1.9 Quality Assurance and Total Quality Management ... 92
- 5.1.10 Organization for Manufacture ... 94

5.2 Definition of CAD/CAM ... 95
5.3 Rational for CAD/CAM ... 98
5.4 Numerical Control and Numerical Control Machine ... 99
- 5.4.1 Numerical Control ... 99
- 5.4.2 Classifications of Numerical Control Machines ... 100

5.5 Solid Freeform Fabrication Methods ... 102
- 5.5.1 Stereolithography ... 102
- 5.5.2 Selective Laser Sintering ... 104
- 5.5.3 Laminated Object Modeling ... 104
- 5.5.4 Fused Deposition Modeling ... 105
- 5.5.5 3D Plotting and Printing Processes ... 106

Part III Production and Processing of Materials

Chapter 6 Powder Metallurgy ... 113

6.1 Introduction ... 113
6.2 Characteristics of Powder Metallurgical Processes ... 114
- 6.2.1 Metal Powders ... 114
- 6.2.2 Powders Produced by Reduction of Ores ... 115
- 6.2.3 Powders Produced by Atomization ... 115
- 6.2.4 Powders Produced by Electrolytic Deposition ... 116

Contents

    6.2.5 Preparation of the Powder ········ 116

Chapter 7 Casting Process and Forming Operation ········ 122

  7.1 Casting Process ········ 122
    7.1.1 Introduction ········ 122
    7.1.2 Sand Casting ········ 124
    7.1.3 Investment Casting ········ 124
    7.1.4 Die Casting ········ 125
  7.2 Forming Operations ········ 127
    7.2.1 Introduction ········ 127
    7.2.2 Forging ········ 127
    7.2.3 Rolling ········ 128
    7.2.4 Extrusion ········ 129
    7.2.5 Drawing ········ 129
    7.2.6 Stamping ········ 130

Chapter 8 Welding ········ 133

  8.1 Introduction ········ 133
  8.2 Definition and Classification of Welding Processes ········ 133
  8.3 Fusion Welding ········ 134
    8.3.1 Shielded Metal Arc Welding ········ 135
    8.3.2 Gas Metal Arc Welding ········ 136
    8.3.3 Gas Tungsten Arc Welding ········ 137
    8.3.4 Plasma Welding ········ 139
    8.3.5 Flux-Cored Wire Welding ········ 140
    8.3.6 Submerged Arc Welding ········ 141
    8.3.7 Electron Beam Welding ········ 142
    8.3.8 Laser Welding ········ 143
  8.4 Welding with Pressure ········ 143
    8.4.1 Resistance Spot Welding ········ 143
    8.4.2 Resistance Seam Welding ········ 144
    8.4.3 Resistance Projection Welding ········ 145
    8.4.4 Cold Pressure Welding ········ 146
    8.4.5 Friction Welding ········ 146
    8.4.6 Explosive Welding ········ 148
    8.4.7 Diffusion Bonding ········ 148
    8.4.8 Magnetically Impelled Arc Butt Welding ········ 149

Chapter 9 Heat Treatment ········ 153

  9.1 Introduction ········ 153

9.2	Full Annealing and Homogenizing	153
9.3	Normalizing and Spheroidizing	157
	9.3.1 Normalizing	157
	9.3.2 Spheroidizing	158
9.4	Structural Changes on Tempering	162
9.5	Thermomechanical Treatments	170
9.6	Surface Hardening	173
	9.6.1 Flame Hardening	173
	9.6.2 Induction Heating	174

Chapter 10　Ceramic Processing Methods ……… 181

- 10.1　Introduction ……… 181
- 10.2　Forming Processing ……… 181
- 10.3　Glass Forming Processing ……… 182
- 10.4　Particulate Forming Processing ……… 183
- 10.5　Drying ……… 185
- 10.6　Firing ……… 185

Chapter 11　Polymer Synthesis ……… 189

- 11.1　Introduction ……… 189
- 11.2　Polyethylene Synthesis ……… 190
- 11.3　Nylon Synthesis ……… 192

Chapter 12　Metal Matrix Composite ……… 198

- 12.1　Introduction ……… 198
- 12.2　Processing of Metal Matrix Composite ……… 202

Chapter 13　Ceramics Matrix Composite ……… 210

- 13.1　Introduction ……… 210
- 13.2　Processing of Ceramic Matrix Composites ……… 215

Chapter 14　Polymer Matrix Composite ……… 223

- 14.1　Introduction ……… 223
- 14.2　Fabrication Processes ……… 226

Chapter 15　Nanocomposite ……… 237

- 15.1　Introduction ……… 237
- 15.2　Organic-Inorganic Nanocomposites ……… 240

Contents

Part Ⅳ Measurement and Analysis of Materials

Chapter 16 Analytical Techniques of Materials ... 249
- 16.1 Introduction ... 249
- 16.2 X-Ray Diffraction ... 249
- 16.3 Optical Metallography ... 253
- 16.4 Scanning Electron Microscopy ... 257
- 16.5 Transmission Electron Microscopy ... 259
- 16.6 High-Resolution Transmission Electron Microscopy ... 261
- 16.7 Scanning Probe Microscopes and Atomic Resolution ... 262
- 16.8 Test of Mechanical Properties ... 266
  - 16.8.1 Loading Methods ... 267
  - 16.8.2 Tensile Strength ... 268
  - 16.8.3 Compressive Strength ... 269
  - 16.8.4 Bend Strength ... 270
- 16.9 Test of Thermal and Electrical Properties ... 272
  - 16.9.1 Thermal Conductivity ... 272
  - 16.9.2 Heat Capacity ... 274
  - 16.9.3 Electrical Resistivity and Conductivity ... 275

Part Ⅴ Application of Materials

Chapter 17 Application of Materials ... 283
- 17.1 Application of Metal Materials ... 283
  - 17.1.1 Used as Structural Materials or Components ... 283
  - 17.1.2 Metals in Electrical and Electronics Applications ... 284
  - 17.1.3 Metals in Aerospace ... 285
  - 17.1.4 On the Road: Metals in Transportation ... 286
  - 17.1.5 Metals and Medicine ... 287
- 17.2 Application of Ceramics Materials ... 288
  - 17.2.1 Refractories ... 288
  - 17.2.2 Uses in the Construction Industry ... 288
  - 17.2.3 Lighting Electrical Appliances ... 289
  - 17.2.4 Electrical Applications ... 289
  - 17.2.5 Communication ... 290
  - 17.2.6 Medical ... 290

  17.2.7 Environmental and Space Applications ················· 291
17.3 Application of Polymeric Materials ······················· 292
  17.3.1 Power of Plastic ·· 292
  17.3.2 Greener Plastics from a Green House Gas ··············· 294
17.4 Application of Composite Materials ······················· 296
  17.4.1 Defense and Military Industry ······························· 296
  17.4.2 Aerospace Vehicle ··· 297
  17.4.3 Transport ·· 299
  17.4.4 Energy Saving and Environment Protect Field ············ 300

Main References ·· 305

Chapter 1 Introduction

1.1 Historical Perspective

The designation of successive historical epochs as the Stone, Copper, Bronze and Iron Ages reflects the importance of materials to mankind. Human destiny and materials resources have been inextricably intertwined since the dawn of history; however, the association of a given material with the age or era that it defines is not only limited to antiquity. The present nuclear and information ages owe their existences to the exploitation of two remarkable elements, uranium and silicon, respectively. Even though modern materials ages are extremely time compressed relative to the ancient metal ages they share a number of common attributes. For one thing, these ages tended to define sharply the material limits of human existence. Stone, copper, bronze and iron meant successively higher standards of living through new or improved agricultural tools, food vessels and weapons. Passage from one age to another was (and is) frequently accompanied by revolutionary, rather than evolutionary, changes in technological endeavors.

It is instructive to appreciate some additional characteristics and implications of these materials ages. For example, imagine that time is frozen at 1 500 BC and we focus on the Middle East, perhaps the world's most intensively excavated region with respect to archaeological remains. In Asia Minor[①] (Turkey) the ancient Hittites[②] were already experimenting with iron, while close by to the east in Mesopotamia[③] (Iraq), the Bronze Age was in flower. To the immediate north in Europe, the south in Palestine[④] and the west in Egypt, peoples were enjoying the benefits of the Copper and Early Bronze Ages. Halfway around the world to the east, the Chinese had already melted iron and demonstrated a remarkable genius for bronze, a copper-tin alloy that is stronger and easier to cast than pure copper. Further to the west on the Iberian Peninsula[⑤] (Spain and Portugal), the Chalcolithic period, an overlapping Stone and Copper Age held sway, and in North Africa

Chapter 1　Introduction

survivals of the Late Stone Age were in evidence. Across the Atlantic Ocean the peoples of the Americas had not yet discovered bronze, but like others around the globe, they fashioned beautiful work in gold, silver and copper, which were found in nature in the free state (*i. e.*, not combined in oxide, sulfide or other ores).

Why materials resources and the skills to work them were so inequitably distributed cannot be addressed here. Clearly, very little technological information diffused or was shared among peoples. Actually, it could not have been otherwise because the working of metals (as well as ceramics) was very much an art that was limited not only by availability of resources, but also by cultural forces. It was indeed a tragedy for the Native Americans, still in the Stone Age three millennia later, when the white man arrived from Europe armed with steel (a hard, strong iron-carbon alloy) guns. These were too much of a match for the inferior stone, wood, and copper weapons arrayed against them. Conquest, colonization and settlement were inevitable. And similar events have occurred elsewhere, at other times, throughout the world. Political expansion, commerce and wars were frequently driven by the desire to control and exploit materials resources, and these continue unabated to the present day.

When the 20th century dawned the number of different materials controllably exploited had, surprisingly, not grown much beyond what was available 2 000 years earlier. A notable exception was steel, which ushered in the Machine Age and revolutionized many facets of life. But then a period ensued in which there was an explosive increase in our understanding of the fundamental nature of materials. The result was the emergence of polymeric (plastic), nuclear, and electronic materials, new roles for metals and ceramics, and the development of reliable ways to process and manufacture useful products from them. Collectively, this modern Age of Materials has permeated the entire world and dwarfed the impact of previous ages.

Only two representative examples of a greater number scattered throughout the book will under score the magnitude of advances made in materials within a historical context. In Figure 1.1, the progress made in increasing the strength-to-density (or weight) ratio⑥ of materials is charted. Two implications of these advances have been improved aircraft design and energy savings in transportation systems. Less visible but no less significant improvements made in abrasive and cutting tool materials are shown in Figure 1.2. The 100-fold⑦ tool cutting speed increase in this century has resulted in efficient machining and manufacturing processes that enable an abundance of goods to be produced at low cost. Together with the dramatic political and social changes in Asia and Europe and the emergence of interconnected global economies, the prospects are excellent that more people will enjoy the fruits of the earth's materials resources than at any other time in history.

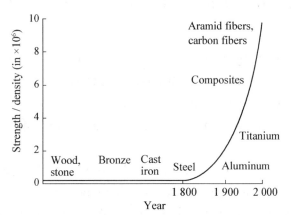

Figure 1.1 Chronological advances in the strength-to-density ratio of materials. Optimum safe load-bearing capacities of structures depend on the strength-to-density ratio. The emergence of aluminum and titanium alloys and, importantly, composites is responsible for the dramatic increase in the 20th century

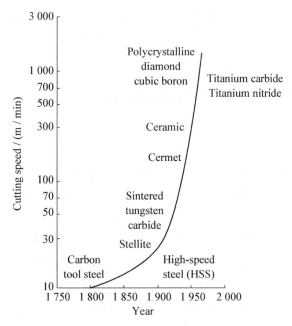

Figure 1.2 Increase in machining speed with the development over time of the indicated cutting tool materials

Chapter 1 Introduction

1.2 Definition of Materials

The materials making up the surrounding world consist of discrete particles, having a submicroscopic size. Their behavior is determined by atomic theories. The states of organization of materials range from the complete disorder of atoms or molecules of a gas under weak pressure to the almost complete order of atoms in a monocrystal.

In this introductory work materials are defined as solids used by man to produce items which constitute the support for his living environment.

Indeed, no object exists without materials. All sectors of human activity depend on materials, from the manufacture of an integrated circuit to the construction of a hydroelectric dam. They appear in our bodies to strengthen or replace our damaged biomaterials. Materials are also as indispensable to our society as food, energy and information. Their essential role is too often forgotten.

The definition employed in this introductory work is limited to solid materials. It excludes liquids and gases, as well as solid combustibles.

1.3 Definition of Engineering

Do you want to change the world? Think engineering! It's everywhere, shaping our world for the better. Engineering is not science. Science is about discovering the natural. Engineering is creating the artificiality. Scientists discover the world that exists; engineers create the world that never was.

Engineering is an incredibly diverse and exciting field. Civil engineering is concerned with making bridges, roads, airports, *etc*. Mechanical engineering deals with the design and manufacture of tools and machines. Electrical engineering is about the generation and distribution of electricity and its many applications. Electronic engineering is concerned with developing components and equipment for communications, computing, and so on. Chemical engineering converts raw materials into usable commodities.

With the rapid advancement of technology, many new fields are gaining prominence and new branches are developing such as computer engineering, soft engineering, nanotechnology, molecular engineering, *etc*. These new specialties sometimes combine with the traditional fields and form new branches such as mechanical engineering and mechatronics, electrical and computer engineering.

Underlying all engineering fields is a common way of thinking, analyzing and solving problems. The way in which engineers think and work has been defined as the engineering method, which has four characteristics.

First, the engineering method is used to analyze, model and solving complex technical problems that require an integrated interdisciplinary view of problem solving.

Another characteristic of the engineering method is that engineers break down complex systems into the simpler components. And in a sense, take the problem apart, not physically but intellectually.

The third characteristic is the development and using of mathematical simulation models. Such models become the engineers' tool for examining, understanding, visualizing and designing complex systems.

Finally, engineering method involves synthesis and design. In this process engineers must take into account factors such as reliability, safety, flexibility, economy and sustainability.

So do you think engineering is right for you? The world today is facing enormous challenges such as global warming, burgeoning population growth, continually increasing energy demands, natural resource depletion and the challenge of sustainable growth. We are convinced that engineers will be able to solve these problems. This country and this world need more young men and women to enter engineering.

1.4 Materials and Engineering

Humankind, materials and engineering have evolved over the passage of time and are continuing to do so. All of us live in a world of dynamic change, and materials are no exception. The advancement of civilization has historically depended on the improvement of materials to work with. Prehistoric humans were restricted to naturally accessible materials such as stone, wood, bones and fur. Over time, they moved from the materials Stone Age into the newer Copper (Bronze) and Iron Ages. Note that this advance did not take place uniformly everywhere, we shall see that this is true in nature even down to the microscopic scale. Even today we are restricted to the materials we can obtain from Earth's crust and atmosphere (Table 1.1). According to Webster's dictionary, materials may be defined as substances of which something is composed or made. Although this definition is

Chapter 1 Introduction

broad, from an engineering application point of view, it covers almost all relevant situations.

Table 1.1 The most common elements in planet Earth's crust and atmosphere by weight percentage and volume

Element	Weight percentage of the Earth's crust
Oxygen (O)	46.60
Silicon (Si)	27.72
Aluminum (Al)	8.13
Iron (Fe)	5.00
Calcium (Ca)	3.63
Sodium (Na)	2.83
Potassium (K)	2.70
Magnesium (Mg)	2.09
Total	98.70
Gas	Percent of dry air by volume
Nitrogen (N_2)	78.08
Oxygen (O_2)	20.95
Argon (Ar)	0.93
Carbon dioxide (CO_2)	0.03

The production and processing of materials into finished goods constitutes a large part of our present economy. Engineers design most manufactured products and the processing systems required for their production. Since products require materials, engineers should be knowledgeable about the internal structure and properties of materials so that they can choose the most suitable ones for each application and develop the best processing methods.

Research and development engineers create new materials or modify the properties of existing ones. Design engineers use existing, modified, or new materials to design and create new products and systems. Sometimes design engineers have a problem in their design that requires a new material to be created by research scientists and engineers. For example, engineers designing a high-speed civil transport (HSCT) (Figure 1.3) will have to develop new high-temperature materials that will withstand temperatures as high as 1 800 ℃ (3 272 ℉) so that airspeeds as high as Mach 12 to 25 can be attained. Research is currently underway to develop new ceramic-matrix composites, refractory intermetallic compounds and single-crystal superalloys[⑧] for this and other similar applications.

Figure 1.3　High-speed civil transport image shows the Hyper-X at Mach 7 with the engines operating. Rings indicate surface flow speeds

One area that demands the most from materials scientists and engineers is space exploration. The design and construction of the international Space Station (ISS) and Mars Exploration Rover (MER) missions are examples of space research and exploration activities that require the absolute best from our materials scientists and engineers. The construction of ISS, a large research laboratory moving at a speed of 27 000 km/h through space, required selection of materials that would function in an environment far different than ours on earth (Figure 1.4). The materials must be lightweight to minimize payload weight during liftoff. The outer shell must protect against the impact of tiny meteoroids and man-made debris. The internal air pressure of roughly 15 psi is constantly stressing the modules. Additionally, the modules must withstand the massive stresses at launch. Materials selection for MERs is also a challenge, especially considering that they must survive an environment in which night temperatures could be as low as $-96\ ℃$. These and other constraints push the limits of material selection in the design of complex systems.

Figure 1.4　The international space station

Chapter 1 Introduction

We must remember that materials usage and engineering designs are constantly changing. This change continues to accelerate. No one can predict the long-term future advances in material design and usage. In 1943 the prediction was made that successful people in the United States would own their own autogyros (auto-airplanes). How wrong that prediction was! At the same time, the transistor, the integrated circuit, and television (color and high definition included) were neglected. Thirty years ago, many people would not have believed that some day computers would become a common household item similar to a telephone or a refrigerator. And today, we still find it hard to believe that some day space travel will be commercialized and we may even colonize Mars. Nevertheless, science and engineering push and transform our most unachievable dreams to reality.

The search for new materials goes on continuously. For example, mechanical engineers search for higher-temperature materials so that jet engines can operate more efficiently. Electrical engineers search for new materials so that electronic devices can operate faster and at higher temperatures. Aerospace engineers search for materials with higher strength-to-weight ratios for aerospace vehicles. Chemical and materials engineers look for more highly corrosion-resistant materials. Various industries look for smart materials and devices and micro-electro-mechanical systems (MEMs) to be used as sensors and actuators in their respective applications. More recently, the field of nanomaterials has attracted a great deal of attention from scientists and engineers all over the world. Novel structural, chemical and mechanical properties of nanomaterials have opened new and exciting possibilities in the application of these materials to a variety of engineering and medical problems. These are only a few examples of the search by engineers and scientists for new and improved materials and processes for a multitude of applications. In many cases, what was impossible yesterday is a reality today!

Engineers in all disciplines should have some basic and applied knowledge of engineering materials so that they will be able to do their work more effectively when using them. The purpose of this book is to serve as an introduction to the internal structure, properties, processing and applications of engineering materials. Because of the enormous amount of information available about engineering materials and due to the limitations of this book, the presentation has had to be selective.

Chapter 1 Introduction

1.5 Selection of Materials

Several requirements must always be studied when selecting the material from which to make a particular component, and the final choice usually involves a compromise. These requirements can be broadly classified as service requirements, fabrication requirements and economic requirements.

1.5.1 Service Requirements

In order that a component be successful in service, it must be of suitable strength, hardness, toughness, elasticity, rigidity, *etc.*, and it must be of a suitable weight. In addition to these basic requirements, certain other properties may also be required, such as suitable electrical, magnetic or thermal properties, heat resistance, and creep or fatigue resistance. Corrosion resistance must usually be as high as possible; alternatively, the material must respond to corrosion resistance treatment.

Even at this stage a compromise is almost always necessary, such as a suitable strength/weight ratio, an acceptable life at high temperature or an acceptable degree of corrosion resistance.

Commercially pure metals have extreme properties; for example, excellent corrosion resistance, but with low strength. It is usual to alloy metals to obtain a suitable compromise between the properties associated with the metals before alloying, or to produce other properties.

The hardness, strength and ductility of some metals and alloys can be modified as a result of fabrication and the component must be designed to suit the manipulation method to be used, and to give the directional properties that are required.

Some alloys will respond to heat treatment so that their strength and hardness can be improved after the manipulation, and it is important that the component be designed to avoid distortion or cracking as a result of this heat treatment.

1.5.2 Fabrication Requirements

It is convenient to classify materials as casting materials and wrought materials. Casting involves heating the material to make it molten and then pouring it into a mould so that it assumes the shape of the impression and retains that shape upon solidification. Wrought materials are suitable for working; working involves the manipulation of the material when it is in the solid state.

Few materials are equally suitable for both casting and working, and very few materials are equally amenable to all variations of casting or working. It is necessary to consider the duty and the shape of the component, and the quantity required and then to select a material with the required properties and which will be suitable for the fabrication method to be used; the component must then be designed to suit the method selected.

The ease with which a material can be machined, and the quality of the finish so obtained is usually very important, as is the suitability of the material for joining by welding, brazing or soldering. As already stated, the properties of some alloys can be altered by heat treatment, but in some cases this treatment is involved and lengthy, and may thus make such materials unsuitable.

1.5.3 Economic Requirements

Except in certain military, research and similar applications, economical considerations are important. These economic considerations should take into account the total cost of the component, which must include the cost of the raw material, and the cost of its manipulation, machining, joining and finishing; the overall costs of a number of materials should be compared, and the use of an expensive material must be justified. The availability of a material will influence its cost, and will in addition affect the delivery date of the components that are made from it; it must be appreciated that the completion date of the product is in itself an economic problem. A material is not necessarily inexpensive because its ore is plentiful; the material cost is often associated with the complexity of the method that must be used to extract it from its ore, or to prepare it for use (Figure 1.5).

1.5.4 Materials That Are Available to the Engineer

Engineering materials can be classified as metallic materials and non-metallic materials.

Although metallic materials are the main ones used in engineering, the non-metallic materials, which include the plastics, polymers, rubber and wood, are of special importance.

The metallic materials are sub-divided① into two groups; these are the ferrous

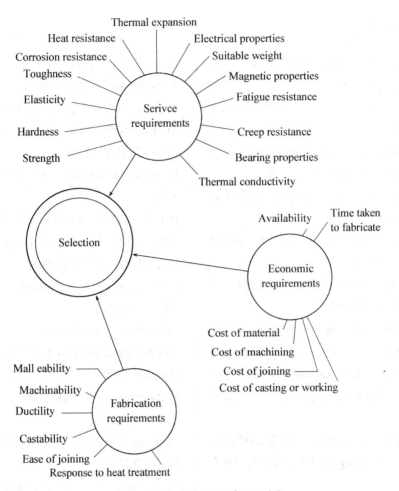

Figure 1.5 Selection of material

alloys, and the non-ferrous metals and alloys. The ferrous alloys contain iron and carbon, to which may be added other elements to confer special properties. The non-ferrous metals are all metals other than iron and its alloys, but the non-ferrous alloys may include small amounts of iron.

Notes

① Asia Minor 小亚细亚
② ancient Hittites 古代赫梯人
③ Mesopotamia 美索不达米亚(亚洲西南部)
④ Palestine 巴勒斯坦
⑤ Iberian Peninsula 伊比利亚半岛
⑥ strength-to-density (or weight) ratio 强度—密度(或重量)比
⑦ 100-fold 100 倍
⑧ single-crystal superalloys 单晶高温合金
⑨ sub-divided 细分;再分

Chapter 1 Introduction

◆ Vocabulary

aerospace [ˈeərəspeɪs] n. 航空宇宙；[航]航空航天空间

antiquity [ænˈtɪkwɪtɪ] n. 高龄；古物；古代的遗物

archaeological [ˌɑːkɪəˈlɒdʒɪkəl] adj. [古]考古学的；[古]考古学上的

autogyros [ˌɔːtəʊˈdʒaɪərəus] n. 旋翼飞机（autogyro 的变形）

ceramics [səˈræmɪks] n. 制陶术，制陶业（ceramic 的复数）

chalcolithic [ˌkælkəˈlɪθɪk] adj. 铜石并用时代的；红铜时代的

colonization [ˌkɒlənaɪˈzeɪʃən] n. 殖民；殖民地化

combustible [kəmˈbʌstəbl] adj. 易燃的；燃烧性的；易激动的 n. 燃质物；可燃物

conquest [ˈkɒŋkwest] n. 征服，战胜；战利品 n. （Conquest）人名；（英）康奎斯特

dramatic [drəˈmætɪk] adj. 戏剧的；引人注目的；激动人心的

ductility [dʌkˈtɪlətɪ] n. 延展性；柔软性；顺从

elasticity [elæˈstɪsɪtɪ] n. 弹性；弹力；灵活性

epoch [ˈiːpɒk] n. [地质]世；新纪元；新时代

flexibility [ˌfleksɪˈbɪlɪtɪ] n. 灵活性；弹性；适应性

hydroelectric [ˌhaɪdrəʊɪˈlektrɪk] adj. 水力发电的；水电治疗的

inequitably [ɪnˈekwɪtəblɪ] adv. 公正地，偏私地

interdisciplinary [ˌɪntəˈdɪsɪplɪn(ə)rɪ] adj. 各学科间的；跨学科的

intermetallic [ˌɪntəmɪˈtælɪk] n. 金属间化合物 adj. 金属间（化合）的

introductory [ˌɪntrəˈdʌkt(ə)rɪ] adj. 引导的，介绍的；开端的

lengthy [ˈleŋθɪ] adj. 漫长的，冗长的；啰唆的

liftoff [ˈlɪftɒf] n. 发射；起飞时刻；搬走

mathematical [mæθ(ə)ˈmætɪk(ə)l] adj. 数学的，数学上的；精确的

mechatronics [mekəˈtrɒnɪks] n. 机电一体化；机械电子学

millennia [mɪˈlenɪə] n. 千年期（millennium 的复数）；一千年；千年庆典；太平盛世

monocrystal [ˈmɒnəˌkrɪstəl] n. 单晶体 adj. 单晶的

payload [ˈpeɪləʊd] n. 净负荷，有效负载，有效载荷

polycrystalline [ˌpɒlɪˈkrɪstəlaɪn] adj. [晶体]多晶的

polymeric [ˌpɒlɪˈmeərɪk] adj. 聚合的；聚合体的

prehistoric [ˌpriːhɪˈstɒrɪk] adj. 史前的；陈旧的

prominence [ˈprɒmɪnəns] n. 突出；显著；突出物；卓越

refrigerator [rɪˈfrɪdʒəreɪtə] n. 冰箱，冷藏库

reliability [rɪlaɪəˈbɪlətɪ] n. 可靠性

rigidity [rɪˈdʒɪdətɪ] n. [物]硬度，[力]刚性；严格，刻板；僵化；坚硬

solidification [səˌlɪdɪfɪˈkeɪʃən] n. 凝固，凝固，固化

sustainability [səˌsteɪnəˈbɪlətɪ] n. 持续

性；永续性；能维持性

tragedy ['trædʒɪdɪ] n. 悲剧；灾难；惨案

uranium [juˈreɪnɪəm] n. [化学]铀

Exercises

1. Translate the following Chinese phrases into English

(1) 铜锡合金　　　　(2) 非金属材料　　　　(3) 自由状态

(4) 制造工艺　　　　(5) 生活环境　　　　　(6) 机械工程

(7) 复杂系统　　　　(8) 人造纤维　　　　　(9) 长远的未来

(10) 材料的选择　　 (11) 服务要求　　　　　(12) 铸造材料

(13) 抗疲劳性　　　 (14) 加工成本　　　　　(15) 铁合金

(16) 非金属材料　　 (17) 材料工程　　　　　(18) 单晶高温合金

2. Translate the following English phrases into Chinese

(1) higher-temperature materials　　(2) corrosion-resistant materials

(3) thermal expansion　　(4) iron-carbon alloy

(5) high-speed steel cutting tool　　(6) solid combustibles

(7) mathematical simulation models　　(8) ceramic-matrix composites

(9) integrated circuit　　(10) fabrication requirements

(11) wrought materials　　(12) non-ferrous metals

(13) heat resistance　　(14) bearing properties

(15) 150-fold　　(16) material science

3. Translate the following Chinese sentences into English

(1) 由此看来，材料的使用完全就是一个选择过程，且此过程又是根据材料的性质从许多的而不是有限的材料中选择一种最适于某种用途的材料。

(2) 但是我们在此的目的不是争辩，而是让大家对材料世界（因为即使是在最遥远的外星球你也能发现相同的元件）和加工领域有一个了解，提供选择材料的方法和工具，以确保上述方面能长期友好地结合起来。

(3) 如果铁轨加工时未采取相应处理，也会发生这样的弯曲。

(4) 有时，材料是否适用于某个应用领域要受到生产和加工操作等方面经济因素的限制。

(5) 说起来很奇怪，最好的导电体像铜和银都不能成为接近绝对零度仍能导电的超导电体。

(6) 纳米材料的研究是一学科跨领域并由物理学家、化学家、材料学家、电机工程师、生物学家和医学家共同参与的研究。

4. Translate the following English sentences into Chinese

(1) The center one is composed of numerous and very small single crystals that are all connected; the boundaries between these small crystals scatter a portion of the light reflected from the printed page, which makes this material optically

translucent.

(2) The more familiar an engineer or scientist is with the various characteristics and structure-property relationships, as well as processing techniques of materials, the more proficient and confident he or she will be to make judicious materials choices based on these criteria.

(3) Materials that are opaque reflect light; those that are transparent refract it, and some have the ability to absorb some wavelengths (colors) while allowing others to pass freely.

(4) This kind of expansion, made without adding heat, is called an adiabatic expansion.

(5) It is the numerous contributions of Materials Science and Technology, which has completely remodeled the world, which supports us by freeing man of a huge number of constraints, linked to our environment.

5. Translate the following Chinese essay into English

产品也经常必须在恶劣的环境中发挥其功能,如暴露于腐蚀性液体、热气体和辐射。像水一样,潮湿的空气也具有腐(浸)蚀性;人手上的汗水特别具有腐(浸)蚀性,当然还有更具腐(浸)蚀性的环境。如果要想产品能够达到其设计使用寿命,所制备的材料就必须能够承受产品运作环境冲击(影响),或者至少要在产品表面上涂覆此类材料。

6. Translate the following English essay into Chinese

One of the reasons that synthetic polymers (including rubber) are so popular as engineering materials lies with their chemical and biological inertness. On the down side, this characteristic is really a liability when it comes to waste disposal. Polymers are not biodegradable, and, as such, they constitute a significant land-fill component; major sources of waste are from packaging, junk automobiles and domestic durables. Biodegradable polymers have been synthesized, but they are relatively expensive to produce. On the other hand, since some polymers are combustible and do not yield appreciable toxic or polluting emissions, they may be disposed of by incineration.

扫一扫,查看更多资料

Part I
Foundation of Material Engineering

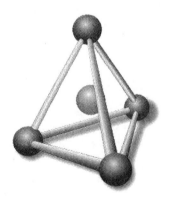

Part 1
Foundation of Material Engineering

Chapter 2　Foundation of Fluid Mechanics

2.1　Characteristics of Flow

The space pervaded the flowing fluid is called flow field, in which the motion parameters are used to describe the flow characteristics, such as velocity u, acceleration a, density ρ, pressure p, temperature T, viscosity force F_v, and so on. The rules of motion parameters which change continuously with the time and the distribution of position for all spatial location (spatial point) occupied by the fluid particles in a flow field are precisely discussed in the fluid dynamics.

2.1.1　A Method to Describe Fluid Motion

Lagrangian viewpoint and Eulerian description are the two main approaches to describe fluid motion.

Lagrange method focuses on the fluid particle or fluid parcel. By watching the trajectory of each individual fluid particle as it travels, the motion parameters of all fluid particles are integrated to get the motion for the fluid field. Lagrange method is seldom used in engineering.

Euler method focuses on the spatial location of flow field. By the observation of motion of fluid particles at some spatial location where the fluid particles pass through, the motion parameters at whole spatial locations are integrated to obtain the motion for the fluid field. Euler method is introduced as follows.

At a definite moment, the velocity of fluid particle at a spatial position is the function of coordinates (x, y, z). The velocity at a fixed point where fluid passes through may be varied at different moment. In other words, the velocity is also a function of time t. Because fluid is continuous medium, the velocity u of a fluid particle is a continuous function of x, y, z and t. It says

$$u = u(x, y, z, t) \qquad (2-1)$$

Chapter 2 Foundation of Fluid Mechanics

Similarly, the pressure p of a fluid particle at a certain spatial point can be expressed as follows

$$p = p(x, y, z, t) \tag{2-2}$$

The acceleration of fluid particle can be obtained from the total differential of the equation (2-3)

$$a = \frac{du}{dt} = u_x \frac{\partial u}{\partial x} + u_y \frac{\partial u}{\partial y} + u_z \frac{\partial u}{\partial z} + \frac{\partial u}{\partial t} \tag{2-3}$$

The acceleration in the equation (2-3) is called substantial derivative or material acceleration. In the component form

$$a_i = \frac{du}{dt} = u_x \frac{\partial u_i}{\partial x} + u_y \frac{\partial u_i}{\partial y} + u_z \frac{\partial u_i}{\partial z} + \frac{\partial u_i}{\partial t} \tag{2-4}$$

in which i is coordinate, $i = x, y, z$; $u_j \partial u_i / \partial u_j (j = x, y, z)$ is the convective acceleration, which represents the difference in velocities of different fluid particles at different spatial position. In other words, for two different spatial points, the difference in the rate or the direction of fluid particles velocity has a contribution to the acceleration. $\partial u_i / \partial t$ is the local acceleration or time-varying acceleration[①], which represents the rate of the velocity of a fluid particle u in a given point with respect to time t in the i direction.

2.1.2 Laminar and Turbulent Flows

Reynolds, a British scientist, explained that there are two basic types of fluid flow, laminar flow and turbulent flow, through an experiment in 1883. In laminar flow, the fluid flows orderly, presenting that one layer of fluid particle moves smoothly over another layer. The particles are not mixed with each other and there is no transverse movement. Laminar flow only shows the pure viscous friction in the fluid layers which is proportional to the fluid shear rate, and the viscous force meets the Newton's law of viscosity, namely, $\tau_v = \mu du/dy$. Turbulent flow shows very complicated and irregular motion trajectories. When the fluid moves along the mainstream direction, there is transverse motion as well; the viscous frictions of turbulent flow consist of not only the viscous friction but also the turbulent shear friction caused by the transverse interchange of momentum in the intermixing of fluid particles.

2.1.3 Steady and Unsteady Flows

Generally speaking, the motion parameters of a fluid flow vary not only with position but also with time. When the motion parameters are dependent on time, the flow is unsteady flow. The flow is steady if the parameters are independent of time.

For unsteady flow, the distributions of velocity and pressure in the flow field are described as the equation (2-1) and the equation (2-2).

For steady flow, these parameters can be expressed as

$$u = u(x, y, z) \tag{2-5}$$
$$p = p(x, y, z) \tag{2-6}$$

Steady flow can be expressed as

$$\begin{aligned} \frac{\partial \rho}{\partial t} &= 0 \\ \frac{\partial \boldsymbol{u}}{\partial t} &= 0 \\ \frac{\partial p}{\partial t} &= 0 \\ \frac{\partial T}{\partial t} &= 0 \end{aligned} \tag{2-7}$$

An example of steady flow is given in Figure 2.1 (a). A water tank consists of a water supply and an overfall to keep the water level at constant. The velocity and pressure in the orifice do not vary with time and the shape of effluent remains a constant jet of flow. Another example is unsteady flow as shown in Figure 2.1 (b). There are no water supply and other attached equipment to keep the water level constant. The water level falls gradually due to drainage in the orifice. The flow in orifice is variable with the water level, in which the velocity and the pressure both vary with time. Most of fluid flow can be treated as steady now in actual engineering.

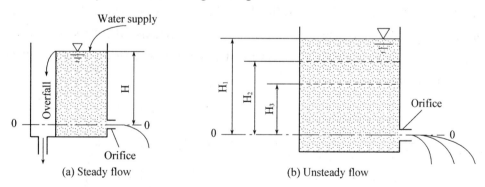

(a) Steady flow (b) Unsteady flow

Figure 2.1 Steady flow and unsteady flow

Chapter 2 Foundation of Fluid Mechanics

2.1.4 Temporal Mean

In turbulent flow, the random movement of fluid particles will lead to fluctuations in the velocity at spatial points. When the fluctuations are small, the flow can be treated as a steady flow by replacing the instantaneous velocity with the mean value in a span of time. The mean value is called temporal mean velocity, denoted by $\bar{u}$

$$\bar{u} = \frac{1}{t}\int_0^t u\,dt \tag{2-8}$$

2.1.5 Uniform and Nonuniform Flows

A uniform flow is the one in which all velocity vectors are identical (in both direction and magnitude) everywhere in a flow field for any given instant. Flow such that the velocity varying from place to place at any instant is nonuniform flow. In uniform flow, there are no convective acceleration, namely, $\partial u_i/\partial j = 0$ (i, j = x, y, z), and therewith the fluid does uniform linear motion. For example, a flow in a straight through tube is the uniform flow and the flow in reducing pipe[②] or winding pipe[③] is nonuniform flow.

2.1.6 Rotational and Irrotational Flows

In the fluid parcel two marked infinitesimal line elements that are at right angles to each other are shown in Figure 2.2. Rotation of a fluid parcel about a given axis, say the z axis, is described as the angular velocity of the two line elements. If the fluid particles are rotational about any axis, the flow is said to be rotational flow, or vortex flow; otherwise it is irrotational flow. Figure 2.2 (b) shows a rotational flow.

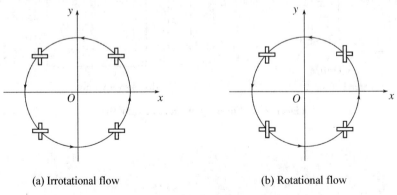

(a) Irrotational flow　　　　　　　　(b) Rotational flow

Figure 2.2　Rotational and irrotational flows

Chapter 2 Foundation of Fluid Mechanics

2.2 Concepts of Flow

2.2.1 Flow Dimensionality

When the variation of the flow parameters transverse to the mainstream direction can be neglected, and all main variables in the flow field can be completely specified by a single coordinate, this flow is one dimensional flow, such as the flow through a long pipeline. Flow in parallel planes is two-dimensional because the motion of fluid particle normal to these planes is restricted for the existence of the planes. Three-dimensional flow is the most general flow.

2.2.2 Path Line and Streamline

A path line or path of particle is the trajectory of an individual fluid particle in flow field during a period of time.

Streamline is a continuous line drawn within fluid flied at a certain instant, the direction of the velocity vector at each point is coincided with the direction of tangent at that point in the line. Streamline vividly indicates the configuration of the fluid motion. One can get any other points in the flow field and trace out numerous streamlines, and then the whole flow field becomes a space permeate with numerous streamlines.

(1) Characteristics of streamline

In unsteady flow, the direction and shape of streamline at a certain point shift from instant to instant as the velocity vector is a variable with time. In steady flow, since there is no change in direction of the velocity vector at any point, the streamline at a certain point keeps steady all along; hence the streamline coincides with the path of particle. Streamline can not be intersected or replicated because a spatial point just has one velocity vector for one instant.

With the help of the concept and characters of streamline, a sketch of fluid flow in various boundary conditions can be visually portrayed in Figure 2.3. From the three flow patterns, it is known that a gently changed solid boundary is the boundary streamline of fluid motion. In other words, the fluid moves along the boundary. If an abrupt change of boundary occurs, the main flow will break away from the boundary due to inertia effect and a vortex area will then come into being.

Chapter 2 Foundation of Fluid Mechanics

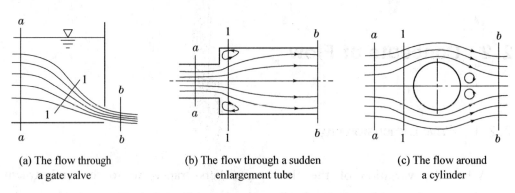

(a) The flow through a gate valve

(b) The flow through a sudden enlargement tube

(c) The flow around a cylinder

Figure 2.3 Flow patterns in various boundary conditions

(2) Differential equation of streamline

Since velocity vector at any point in the streamline is tangent to streamline, the streamline in differential form is given as

$$\frac{\mathrm{d}x}{u_x} = \frac{\mathrm{d}y}{u_y} = \frac{\mathrm{d}z}{u_z} \qquad (2-9)$$

For example, the velocity components in a known plane flow field are

$$u_x = -\frac{ky}{x^2 + y^2}$$
$$u_y = \frac{kx}{x^2 + y^2} \qquad (2-10)$$
$$u_z = 0$$

Determine the streamline.

From the equation (2-9), the velocity components are given

$$-\frac{\mathrm{d}x}{y} = \frac{\mathrm{d}y}{x} \qquad (2-11)$$

making a integral

$$x^2 + y^2 = C \qquad (2-12)$$

Various streamlines can be obtained by assigning different constant C and the streamlines are a cluster of circles in various radii. The flow is steady flow because the given velocity is independent of the time.

2.2.3 Stream Tube, Ministream Tube and Total Flow

(1) Stream tube

The stream tube is composed of the streamlines passing through every point in a closed curve (not the streamline) C_0 which is drawn in the flow field. An example of stream tube is shown in Figure 2.4.

Chapter 2 Foundation of Fluid Mechanics

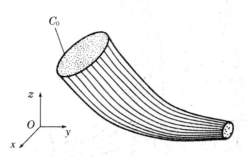

Figure 2.4 Stream tube

The shape of stream tube changes with time in unsteady flow and remains unchanging in steady flow. Since the surface of stream tube is formed with streamlines, the fluid can not penetrate through the surface of stream tube. For this reason, the stream tube is likely a solid wall and the flow is restricted in the tube. In steady flow the stream tube is the same with an actual pipe.

(2) Ministream tube

The stream tube with an infinitesimal section is said to be ministream tube. Streamline is the extreme case of ministream tube.

(3) Total flow

Total of countless ministream tubes is called total flow.

In the analyses of the variation of motion parameters such as flow velocity flow rate, pressure, and so on, the total flow can be divided into countless ministream tube. The motion parameters at every point on section dA can be considered to be uniform due to the very small section of ministream tube, therefore, the flow parameters of total flow can be obtained by the integral method.

2.2.4 Cross Section, Flow Rate and Average Velocity

(1) Flow section

The flow section is a section that every area element in the section is normal to ministream tube or streamline. The flow section is denoted as dA or A. Two flow sections I and II are given in Figure 2.5. In the section I streamlines are curvous and the flow section is a curved surface. In the section II, streamlines are all the parallel beelines, and the flow section is a plane. In actual application, the flow section should be selected to make the streamlines parallel, hence, the flow section is a plane area, or a cross section such as the sections II, 1-1 and 2-2 as shown in Figure 2.5.

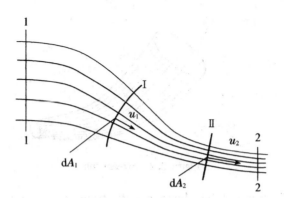

Figure 2.5 Flow section

(2) Flow rate

The amount of fluid passing through a cross section in unit interval is called flow rate or discharge. The discharge usually is represented by volume, mass and weight of the fluid and accordingly named as the volumetric flow rate (also known as volume flow rate) Q, the rate of mass flow (mass flow rate) M, and the rate of weight flow (weight flow rate) G. The relationships of them are

$$G = \rho g Q \text{ (N/s)} \tag{2-13}$$

$$M = \frac{\rho}{g} Q \text{ (kg/s)} \tag{2-14}$$

$$Q = \frac{M}{\rho} = \frac{G}{\gamma} \text{ (m}^3\text{/s)} \tag{2-15}$$

Without special instruction, the volumetric flow rate is referred to as flow rate or discharge.

For a ministream tube, the volumetric flow rate dQ is equal to product of flow velocity u and element of cross section dA, namely

$$dQ = u dA \tag{2-16}$$

For a total flow, volumetric flow rate Q can be obtained by integration of the flow rate dQ in a ministream tube over the cross section A, namely

$$Q = \int_A u \, dA \tag{2-17}$$

(3) Average velocity

The velocity at every point in the cross section is different and its distribution is a parabola. The velocity u takes the maximum u_{max} on the pipe axle and the zero on the boundary as shown in Figure 2.6. According to the equivalency of flow rate, the fluid volume that pass through a cross section A as an average velocity V must be equal to the fluid volume passing through the section as an actual velocity in unit time, namely, $VA = \int_A u \, dA = Q$, therewith

$$V = \frac{\int_A u\,dA}{A} = \frac{Q}{A} \qquad (2-18)$$

The average velocity according to the equivalency of flow rate is called the section average velocity. The velocity of a pipeline in engineering is the section average velocity V.

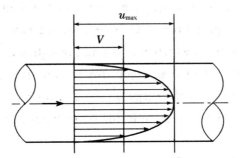

Figure 2.6　Distribution of velocity over cross section

2.3　Reynolds Transport Equation

2.3.1　System

In fluid mechanics, a system is a set of definite fluid particles selected in the interest of researcher. System boundary is the surroundings that distinguish the set from the other material and named it the surface of system. The surface moves with the fluid and changes with time. There is no exchange of the mass but the energy on boundary. The boundary suffers surface forces from surroundings. The dynamics of a system can be expressed by Newton's second law of motion as

$$\sum F = \frac{d(mu)}{dt} \qquad (2-19)$$

in which $\sum F$ is the resultant of all external forces acting on the system; m is the mass of the definite particles, a fixed value, and u is the velocity at the center of mass of the system.

2.3.2　Control Volume

In Eulerian method of analysis the control volume (cv) is defined as an invariably hollow volume or an invariably hollow frame. The term "invariably" means that the shape, volume and flow areas of control volume have no change with time as the control volume is chosen. The boundary of control volume is called control surface (**cs**). In the control surface the area where the fluid particles may come in and out is called the flow control surface and the others are the no-flow control surfaces[④], which usually are the solid walls. Control volume is stationary in the coordinate system, in which the points can be expressed by coordinates of the coordinate system. The spatial points including the interior points of boundary are occupied by the fluid particles which are passing through and the particles do not the same at the different instant of time except for the relative static flow field. There may be the exchange of mass and energy on the control surfaces which can be acted by distributed forces from the medium in the outer boundary of them.

2.3.3 Reynolds Transport Equation

Considering a flow field as shown in Figure 2.7, the solid line volume τ is the control volume, noted by **cv**; the dotted-line volume is the system, noted by $N(t)$, whose property is a term of a general property of the system, such as mass, momentum, and so on. The flow situation at time t_1 ($t_1 > t$) is shown in Figure 2.7(b), in which the volume τ_3 and τ_1 are **cv**, and volume τ_1 and τ_2 are the system $N(t_1)$. The control surface A contains inflow area A_i and outflow area A_0. The velocity of fluid particle at a spatial point in **cv** at time t is represented by $u(r, t)$, in which r is radius vector of spatial point relative to origin of coordinates, $r = r(x, y, z)$. The general property per unit mass is $\eta(r, t)$. At time t_1, the system $N(t)$ moves to the dotted-line position in Figure 2.7(b) and becomes $N(t_1)$, the velocity $u(r, t_1)$ and the general property per unit mass $\eta(r, t_1)$.

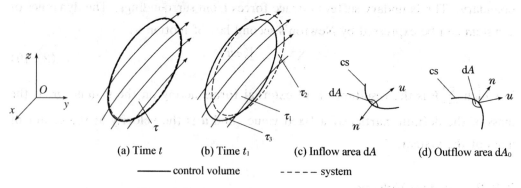

Figure 2.7 A system flow through control volume

The increment of system property $N(t)$ in time $(t_1 - t)$ is

$$N(t_1) - N(t) = \iiint_{\tau_1 + \tau_2} \rho \eta(\boldsymbol{r}, t_1) d\tau - \iiint_{\tau_1 + \tau_3} \rho \eta(\boldsymbol{r}, t) d\tau \qquad (2-20)$$

Now considering the rate of increment in the system $N(t)$ as the time t_1 approaching to time t

$$\frac{dN(t)}{dt} = \lim_{t_1 \to t} \frac{N(t_1) - N(t)}{t_1 - t} = \lim_{t_1 \to t} \left[\iiint_{\tau_1 + \tau_2} \rho \eta(\boldsymbol{r}, t_1) d\tau - \iiint_{\tau_1 + \tau_3} \rho \eta(\boldsymbol{r}, t) d\tau \right] \qquad (2-21)$$

$$\frac{dN(t)}{dt} = \lim_{t_1 \to t} \frac{1}{t_1 - t} \left\{ \iiint_{\tau_1} \rho [\eta(\boldsymbol{r}, t_1) - \eta(\boldsymbol{r}, t)] d\tau \right\} +$$

$$\lim_{t_1 \to t} \frac{1}{t_1 - t} \left[\iiint_{\tau_2} \rho \eta(\boldsymbol{r}, t_1) d\tau - \iiint_{\tau_3} \rho \eta(\boldsymbol{r}, t) d\tau \right] \qquad (2-22)$$

From Lagrange mean value theorem[5] having

$$\eta(\boldsymbol{r}, t_1) - \eta(\boldsymbol{r}, t) = \frac{\partial \eta(\boldsymbol{r}, \xi)}{\partial t} \bigg|_{\xi = t + \theta(t_1 - t)} (t_1 - t) \qquad (2-23)$$

in which $0 \leqslant \theta \leqslant 1$.

When $(t_1 - t)$ comes near zero, volume τ_1 approaches to volume τ. Volume τ is a constant volume as it can learn from the definition of **cv**. So

$$\lim_{t_1 \to t} \frac{1}{t_1 - t} \left\{ \iiint_{\tau_1} \rho [\eta(\boldsymbol{r}, t_1) - \eta(\boldsymbol{r}, t)] d\tau \right\} = \iiint_{\tau} \rho \frac{\partial \eta(\boldsymbol{r}, t)}{\partial t} d\tau \qquad (2-24)$$

and

$$\lim_{t_1 \to t} \frac{1}{t_1 - t} \left\{ \iiint_{\tau_2} \rho \eta(\boldsymbol{r}, t_1) d\tau \right\} = \lim_{t_1 \to t} \frac{1}{t_1 - t} \iint_{A_0} \rho \eta(\boldsymbol{r}, t_1) \boldsymbol{u}(\boldsymbol{r}, t_1)(t_1 - t) dA_0$$

$$= \iint_{A_0} \rho \eta(\boldsymbol{r}, t_1) \boldsymbol{u}(\boldsymbol{r}, t) dA_0 \qquad (2-25)$$

in which dA_0 is the product of the outflow area element dA_0 and the outer normal unit vector $\boldsymbol{n}$

In the same way

$$\lim_{t_1 \to t} \frac{1}{t_1 - t} \left\{ \iiint_{\tau_3} \rho \eta(\boldsymbol{r}, t_1) d\tau \right\} = - \iint_{A_i} \rho \eta(\boldsymbol{r}, t) \boldsymbol{u}(\boldsymbol{r}, t) dA_i \qquad (2-26)$$

in which the minus sign indicates that the outer normal unit vector $\boldsymbol{n}$ of inflow area is opposition to the direction of flow velocity $\boldsymbol{u}$.

Let $A = A_i + A_0$, So

$$\iint_{A_0} \rho \eta(\boldsymbol{r}, t) \boldsymbol{u}(\boldsymbol{r}, t) dA_0 - \left(- \iint_{A_i} \rho \eta(\boldsymbol{r}, t) \boldsymbol{u}(\boldsymbol{r}, t) dA_i \right) = \iint_A \rho \eta(\boldsymbol{r}, t) \boldsymbol{u}(\boldsymbol{r}, t) dA$$

$$(2-27)$$

The equation (2-27) and the equation (2-24) may combined into the equation (2-25) and it gives

Chapter 2 Foundation of Fluid Mechanics

$$\frac{dN(t)}{dt} = \iiint_\tau \frac{\partial[\rho\eta(r,t)]}{\partial t} d\tau + \iint_A \rho\eta(r,t)u(r,t) dA \qquad (2-28)$$

The equation is Reynolds transport equation.

2.3.4 Application of the Reynolds Transport Equation

In the last section, the Reynolds transport equation is obtained by taking the superposition of a system and a control volume. When general property is given in a definite physical quantity of flow field, the specific form of conservation of mass, energy conservation and momentum equation in the fluid flow field can be derived.

(1) Continuity equation

No matter how the shape of system may be, the mass remains constant in a system with definite particles according to the principle of conservation of mass, i.e.

$$\frac{dm}{dt} = 0 \qquad (2-29)$$

Let the general property N be the mass of system m. Then the general property per unit mass η is 1. Noticing that the control volume is invariable in its volume, from the equation (2-24) the continuity equation can be expressed as

$$0 = \frac{\partial}{\partial t}\iiint_\tau \rho d\tau + \iint_A \rho u dA \qquad (2-30)$$

For a steady flow, $\partial\left(\iiint_\tau \rho d\tau\right)/\partial t = 0$, so

$$\iint_A \rho u \cdot dA = 0 \qquad (2-31)$$

namely, the net rate of mass inflow and outflow to the control volume is zero.

(2) Equation of momentum

Let $N = mu$, then momentum per unit mass is v, by the use of the equation (2-28)

$$\sum F = \frac{d(mu)}{dt} = \frac{\partial}{\partial t}\iiint_\tau \rho u d\tau + \iint_A (\rho u)u \cdot dA \qquad (2-32)$$

In other words, the resultant external force acting on a control volume is equal to the sum of variation of the momentum within the control volume plus the net outflow of momentum from the control volume in unit time.

(3) Energy equation

The law of conservation of energy is expressed as that the input heat Q_H of a system minus the work W done by the system is equal to the net variation of energy in the system when the system experiences a process from the initial state to the final state.

$$Q_H - W = N_2 - N_1 \tag{2-33}$$

in which N_1, N_2 are energy in the initial and final states of system, respectively.

Taking the general property N as the total energy of system which contains internal energy E_i and kinetic energy $mu^2/2$, and then general property per unit mass is $\eta = e_i + u^2/2$, from the equation (2-28), the energy conservation equation can be expressed as

$$\frac{\delta Q_H}{\delta t} - \frac{\delta W}{\delta t} = \frac{dN}{dt} = \frac{\partial}{\partial t}\iiint_\tau \rho\left(e_i + \frac{u^2}{2}\right)d\tau + \iint_A \rho\left(e_i + \frac{u^2}{2}\right)\boldsymbol{u}\cdot d\boldsymbol{A} \tag{2-34}$$

in which δQ_H is the difference of heat of the system from initial to final state; δW is the difference of work done by the system from initial to final state; u is the velocity of fluid particle on dA; e_i is the internal energy per unit mass of a fluid, which depends on the temperature of fluid, $de_i = c_V dT$, in which T is the absolute temperature, K; c_V is the specific heat capacity at constant volume and c_V is a constant for liquid, but $c_V = f(T)$ for gas.

Neglecting nuclear energy, magnetic energy and surface tension effect and supposing that the mass force is potential and only in the gravity, the work W done by a system may be broken into three parts: the work W_p done by pressure, the work W_s done by shear force and the work W_g done by mass force. In time $(t_1 - t)$

$$\delta W = \delta W_p + \delta W_s + \delta W_g \tag{2-35}$$

in which

$$\delta W_p = \delta t \int_A p\boldsymbol{u}\cdot d\boldsymbol{A} \tag{2-36}$$

$$\delta W_g = \delta t \int (gz)\boldsymbol{u}\cdot d\boldsymbol{A} \tag{2-37}$$

Hence

$$\frac{\delta Q_H}{\delta t} - \frac{\delta W_s}{\delta t} = \frac{\partial}{\partial t}\iiint_\tau \rho\left(e_i + \frac{u^2}{2}\right)d\tau + \iint_A \left(\frac{p}{\rho} + gz + e_i + \frac{u^2}{2}\right)\rho\boldsymbol{u}\cdot d\boldsymbol{A} \tag{2-38}$$

For steady flow

$$\frac{\partial}{\partial t}\iiint_\tau \rho\left(e_i + \frac{u^2}{2}\right)d\tau = 0 \tag{2-39}$$

The shear forces act on the surrounding of a system. For example, the shear forces on the blade are given by the torque exerted on a rotating spindle in a turbine. If there are no energy input and output devices such as pump or turbomachine in the system, $\delta W_s/\delta t = 0$, therewith

$$\frac{\delta Q_H}{\delta t} = \iint_A \left(\frac{p}{\rho} + gz + e_i + \frac{u^2}{2}\right)\rho\boldsymbol{u}\cdot d\boldsymbol{A} \tag{2-40}$$

If the flow is adiabatic, there is $\delta Q_H/\delta t = 0$; and then

$$\iint_A \left(\frac{p}{\rho} + gz + e_i + \frac{u^2}{2}\right)\rho\boldsymbol{u}\cdot d\boldsymbol{A} \tag{2-41}$$

Soon

$$gz + \frac{u^2}{2} + \frac{p}{\rho} + e_i = C \qquad (2-42)$$

The more discussions for energy equation are presented in Section 2.6.

2.4 Continuity Equation

As fluid is continuous medium, it is thought that the whole space is occupied fully and continuously with the fluid, and without a point source or a point sink when the fluid motion is dealt. This is the continuous condition of fluid motion. According to the rule of conservation of mass, for a closed surface (i.e. control volume) fixed in the space, the difference in fluid mass of inflow and outflow should equal to the variation of fluid mass in the closed surface for an unsteady flow. When the flow is steady, the fluid mass of inflow must be equal to that of outflow. If these conclusions are expressed as a mathematic form, it is the equation of continuity.

2.4.1 Steady Flow Continuity Equation of One-Dimensional Ministream Tube

A stream tube is shown in a flow field of Figure 2.8, the flow sections are A_1 and A_2, and the other section is no flow section. In the stream tube, a ministream tube, with flow sections dA_1, dA_2; velocities u_1, u_2 and densities ρ_1, ρ_2, is selected. There are inflow and outflow of the fluid on the surface dA_1 and dA_2 is the ministream tube. The other surface is surrounded by stream lines and no fluid can get across it.

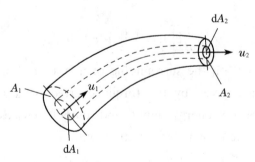

Figure 2.8 One-dimensional stream tube

In time dt the mass inflow at dA_1 is $\rho_1 u_1 dA_1 dt$ and the mass outflow at dA_2 is $\rho_2 u_2 dA_2 dt$. The net mass inflow of fluid for the ministream tube is

$$dM = (\rho_1 u_1 dA_1 - \rho_2 u_2 dA_2) dt \qquad (2-43)$$

If the flow is steady, the shape of ministream tube and the motion parameters at any point do not change with time. Because the fluid is continuous medium, the fluid mass enclosed within the sections dA_1 and dA_2 in time dt does not change. So

$$dM = 0 \qquad (2-44)$$

then

$$\rho_1 u_1 dA_1 = \rho_2 u_2 dA_2 \qquad (2-45)$$

The formula is the equation of continuity for compressible fluid, steady flow along with a ministream tube or stream line.

If the fluid is incompressible, $\rho_1 = \rho_2 = \rho$, hence

$$u_1 dA_1 = u_2 dA_2 \qquad (2-46)$$

The formula is the continuity equation for incompressible fluid, steady flow along with ministream tube. By the way, the formula is suitable to the unsteady low too.

2.4.2 Total Flow Continuity Equation for One-Dimensional Steady Flow

The integrals are made at both sides of the equation (2-44) over the sections A_1 and A_2

$$\int_{A_1} \rho_1 u_1 dA_1 = \int_{A_2} \rho_2 u_2 dA_2 \qquad (2-47)$$

In above equation, if the ρ_1 and ρ_2 are replaced by the average density ρ_{1m} and ρ_{2m} over the sections, it can be written as

$$\rho_{1m} \int_{A_1} u_1 dA_1 = \rho_{2m} \int_{A_2} u_2 dA_2 \qquad (2-48)$$

integrating and having

$$\rho_{1m} Q_1 = \rho_{2m} Q_2 \qquad (2-49)$$

or

$$\rho_{1m} V_1 A_1 = \rho_{2m} V_2 A_2 \qquad (2-50)$$

in which V_1, V_2 are respectively the average velocities of fluid over the sections A_1 and A_2, m/s; A_1, A_2 are respectively the areas of the section of stream tube, m².

The equation (2-50) can also be gotten directly from the equation (2-30). This equation shows that the rate of mass flow remains unchanged along the flow path for the compressible fluid in a steady flow.

For incompressible fluid flow, the density ρ is a constant. The equation (2-50) reduces to

$$Q_1 = Q_2 \qquad (2-51)$$

or

$$V_1 A_1 = V_2 A_2 \qquad (2-52)$$

The equation (2 - 52) is the total flow continuity equation for the incompressible fluid in steady flow. The equation states that volumetric flow rate of one-dimensional total flow remains a constant along the flow path under the situation of steady flow. It also can be seen that the average velocity varies inversely with the area of cross section, in other words, a big area of cross section has the low velocity and vice versa. By the way, the equation (2 - 52) can be used to the incompressible fluid in an unsteady flow.

2.4.3 Three-Dimensional Differential Continuity Equation

In a flow field, a hexahedron with the sides dx, dy and dz is selected as the control volume as shown in Figure 2.9. Suppose the velocity components of fluid particles in the x, y, z directions at the point $m(x, y, z)$ in the hexahedron are u_x, u_y, u_z, respectively, and the density of fluid is ρ. Within unit time, the inflow of mass through the left face on the hexahedron is $\rho u_y dxdz$, and accordingly the outflow of mass from the right face of hexahedron is $\{\rho u_y + [\partial(\rho u_y)/\partial y]dx\}dxdz$. In time dt, the net mass flux outflow in the y direction through these two faces is

$$\left[\rho u_y + \frac{\partial(\rho u_y)}{\partial y}dy\right]dxdzdt - \rho u_y dxdzdt = \frac{\partial(\rho u_y)}{\partial y}dxdydzdt \qquad (2-53)$$

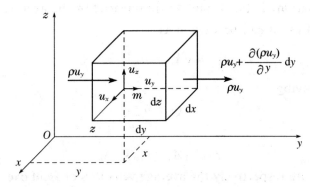

Figure 2.9 A hexahedron as control volume

In the same way, in time dt the net mass flux in the x and z direction yield $[\partial(\rho u_x)/\partial x]dxdydzdt$ and $[\partial(\rho u_z)/\partial z]dxdydzdt$, respectively. Hence, in time dt, the net mass flux outflow for the whole hexahedron is

$$\left[\frac{\partial(\rho u_x)}{\partial x} + \frac{\partial(\rho u_y)}{\partial y} + \frac{\partial(\rho u_z)}{\partial z}\right]dxdydzdt \qquad (2-54)$$

At the beginning of time dt, the fluid density at point m is ρ, and at the end time of dt the density at the point becomes $\rho + (\partial\rho/\partial t)dt$. The total reduction in mass aroused by the variation in density is

Chapter 2 Foundation of Fluid Mechanics

$$\rho dxdydz - \left(\rho + \frac{\partial \rho}{\partial t}dt\right)dxdydz = -\frac{\partial \rho}{\partial t}dxdydzdt \qquad (2-55)$$

The fluid flow is continuous and there are no source or sink in the hexahedron. According to the law of conservation of mass, in time dt, the reduction in mass must be equal to the difference in mass between the outflow and the inflow, i.e.,

$$\left[\frac{\partial(\rho u_x)}{\partial x} + \frac{\partial(\rho u_y)}{\partial y}\frac{\partial(\rho u_z)}{\partial z}\right]dxdydzdt = -\frac{\partial \rho}{\partial t}dxdydzdt \qquad (2-56)$$

Simplifying

$$\frac{\partial \rho}{\partial t} + \frac{\partial(\rho u_x)}{\partial x} + \frac{\partial(\rho u_y)}{\partial y}\frac{\partial(\rho u_z)}{\partial z} = 0 \qquad (2-57)$$

The equation (2 - 57) is exactly a differential expression of the continuity equation for compressible fluid flow. The physical meaning of the equation is that in a unit time the algebraic sum of net mass flux per unit volume between the outflow and inflow and the variation of mass within the volume is zero. This equation is actually an embodiment of the law of conservation of mass in fluid mechanics.

For compressible fluid and steady flow, $\partial \rho / \partial t = 0$, the equation (2 - 57) becomes

$$\frac{\partial(\rho u_x)}{\partial x} + \frac{\partial(\rho u_y)}{\partial y}\frac{\partial(\rho u_z)}{\partial z} = 0 \qquad (2-58)$$

The equation (2 - 58) is the three-dimensional continuity equation for the compressible fluid in steady flow.

For incompressible fluid, ρ is constant, the equation (2 - 57) becomes

$$\frac{\partial u_x}{\partial x} + \frac{\partial u_y}{\partial y}\frac{\partial u_z}{\partial z} = 0 \qquad (2-59)$$

The equation (2 - 59) can be applied to both steady flow and unsteady flow. $\partial u_x / \partial x$ represents the rate of linear deformation of fluid particle in the x direction. Therefore, the equation (2 - 59) indicates the sum of the rate of linear deformation of fluid particle in the three directions is zero, namely, if there is the stretch in one direction, absolutely there exists the contraction in other direction.

2.5 Bernoulli Equation

2.5.1 Euler's Equation in One-Dimensional Flow for Ideal Fluid

In Figure 2.10 a control volume of cylindrical element, with flow area dA and length ds, in a streamline is selected. In the flow field that the mass force is only the gravity, the motion of fluid along a streamline is discussed. The forces acting on the control volume are the gravity force and pressure forces on the both of end surfaces. In the tangent direction, the gravity force is $\rho g ds dA \cos\theta$ and the pressure on the outflow section is $p + (\partial p/\partial s)\delta s$.

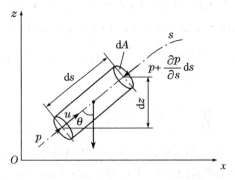

Figure 2.10 A cylinder as control volume

The resultant forces acting on the cylinder can be expressed as

$$F_s = p\delta A - \left(p\delta A + \frac{\partial p}{\partial s}\delta s \delta A\right) - \rho g \delta s \delta A \cos\theta = -\frac{\partial p}{\partial s}\delta s \delta A - \rho g \delta s \delta A \cos\theta \tag{2-60}$$

According to the Newton's second law of motion

$$F_s = \rho \delta s \delta A \frac{du}{dt} \tag{2-61}$$

in which du/dt is the acceleration of the element of fluid.

From the definition of material acceleration for a fluid particle it has

$$\frac{du}{dt} = u\frac{\partial u}{\partial s} + \frac{\partial u}{\partial t} \tag{2-62}$$

From the geometric relationships in Figure 2.10 $\cos\theta = \partial z/\partial s$, the equation reduces to

$$\frac{\partial u}{\partial t} + u\frac{\partial u}{\partial s} = -\frac{1}{\rho}\frac{\partial p}{\partial s} - g\frac{\partial z}{\partial s} \tag{2-63}$$

For steady flow, $\partial u/\partial t = 0$, s is therefore the only variable and a total differential may replace the partials

$$udu + \frac{dp}{\rho} + gdz = 0 \tag{2-64}$$

This is Euler's equation in differential form.

2.5.2 Bernoulli's Equation

Euler's equation in differential form given in the equation (2-64) is used in the situations that the ideal fluid flows along a streamline in steady. Because the ideal fluid is incompressible, the Bernoulli's equation can be obtained with an integral along a streamline as follows

$$\frac{u^2}{2} + \frac{p}{\rho} + gz = C(m^2/s^2) \qquad (2-65)$$

From the process of derivation for the equation (2-64), it can be seen that the equation (2-65) is an energy equation per unit mass. It has the dimensions $(L/T)^2$ because $m \cdot N/kg = (m \cdot kg \cdot m/s^2)/kg = m^2/s^2$. So the meanings for each term in the Bernoulli's equation are as follows:

The first term is the kinetic energy per unit mass and it shows that for a mass of fluid m the kinetic energy per unit mass is $(mu^2/2)/m$, i.e., $u^2/2$.

The second term is the pressure energy per unit mass. It is interpreted as the work done by the pressure force pdA which move to the distance ds along the streamline.

The third term is the potential energy per unit mass. It indicates the work needed to lift weight, $\rho ds dA g$ an elevation z.

The sum of the three terms, kinetic energy, pressure energy and potential energy, is the mechanical energy of fluid. The equation (2-65) shows that the total mechanical energy per unit mass of fluid remains constant at any position along the flow path but the three types of energy might be transformed each other in the path. Therefore, Bernoulli's equation for the ideal fluid actually is a representation of the transformation and conservation law of energy in fluid mechanics.

Multiplying equation (2-65) by density ρ, the Bernoulli's equation per unit volume is given

$$\rho \frac{u^2}{2} + p + \rho gz = C_1 (N/m^2) \qquad (2-66)$$

which is convenient for the calculation of gas flow. Because the dimension of $\rho u^2/2$ is the same as that of pressure, it is called dynamic pressure.

Since the specific weight ρg is very small for gas, the term ρgz can be eliminated and the equation (2-66) reduces to

$$\rho \frac{u^2}{2} + p = C_2 (N/m^2) \qquad (2-67)$$

If each term in the equation (2-65) is divided by the acceleration of gravity g, the Bernoulli's equation per unit weight can be obtained as

$$\frac{u^2}{2g} + \frac{p}{\rho g} + z = C_2 (m \cdot N/N, \text{ or } m) \qquad (2-68)$$

For arbitrary two points 1 and 2 along a streamline, the equation (2-68) could be written as

$$\frac{u_1^2}{2g} + \frac{p_1}{\rho g} + z_1 = \frac{u_2^2}{2g} + \frac{p_2}{\rho g} + z_2 \qquad (2-69)$$

Bernoulli's equation offers a relationship between the motion variables p, u, ρ and the coordinate z along a streamline. When the fluid is in static state, $u = 0$, the equation (2-68) reduces to

$$\frac{p}{\rho g} + z = C_3 \; (\text{m} \cdot \text{N/N, or m}) \tag{2-70}$$

This is the basic equation of hydrostatics which is a special case of Bernoulli's equation.

2.5.3 Geometry Expression of Bernoulli's Equation

Bernoulli's equation of the equation (2-68) in unit weight has a dimension of length, and the transformation of the three types of energy along a streamline in the flow process could be expressed in a graph of geometry. The dimension of z is the length; it refers the elevation of the fluid particle to lift from the datum plane and is called the elevation head. The dimension of $p/(\rho g)$ is also the length, it refers the elevation that the fluid particle under the pressure p can reach at from the position z. The elevation is called pressure head. The dimension of $u^2/(2g)$ is $L^2 T^{-2}/(LT^{-2}) = L$, it is the dimension of length too. It refers what height can be reached at for the fluid particle that is sprayed upward with the velocity u (neglecting air friction) and is called velocity head, denoted as h_u as shown in Figure 2.11.

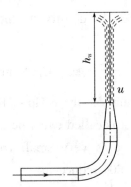

Figure 2.11 Velocity head h_u

The sum of elevation head, pressure head and velocity head is called the total head in Bernoulli's equation and is expressed as H. Because of each team of Bernoulli's equation represents an elevation; the relationship of them could be indicated in a geometric graph. In Figure 2.12, the datum line for the elevation is 0-0; the line segment *amb* (dash-dotted line[6]) is the line which connects each point z, it is called elevation head line, the segment *enf* (solid line) is the line which connected each vertex point of $z + p/(\rho g)$, it is called the piezometer head line; the segment *gkh* is the line which connected each vertex point of $u^2/(2g)$, it is called the total head line. The meaning of Bernoulli's equation for ideal fluid in geometry is as follows: The total head line is a horizontal line, each head could increase or decrease but the total head remains constant. In Figure 2.12, the total heads at the three points a, m and b are $H_1 = z_1 + p_1/(\rho g) + u_1^2/(2g)$ and $H_2 = z_2 + p_2/(\rho g) + u_2^2/(2g)$. Three total heads are all equal, namely

$$H_1 = H = H_2 \qquad (2-71)$$

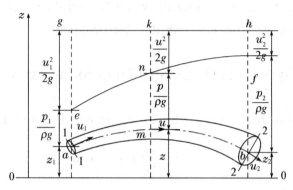

gkh-total head line; *enf*-piezometer head line; *amb*-elevation head line; 0-0-datum line

Figure 2.12 Total head line and piezometer head line

2.6 Flow Losses and Steady Flow Energy Equation

2.6.1 Reversible and Irreversible Processes

A process may be defined as the evolution of a system proceeding from a state to another in succession. The process could cause some change of a system and its surroundings. If the system could be reverted to the original state with the elimination of effect on the surrounding caused in the former process, this process is reversible. In other words, for an original state of system A and a state of environment B, when the system is undergoing a process, the state of system becomes A′ and the state of environment becomes B′. If there is another process, in which the state of system could go back to A and the state of environment back to B, the former process is called a reversible process. Conversely, if there is no way to make the system and the surroundings to recover, the process is irreversible.

The irreversible process is universal due to the ineluctability of friction. For example, mechanical energy losses caused by the resistance of friction in the flow are transformed to heat energy, and then dissipated away. This process is irreversible. For ideal fluid that the viscosity is not taken into accounted, the process of energy exchange in which the fluid system does work to surrounding is reversible. For example, in the absence of resistance due to friction the potential energy in a high level reservoir is transformed to the kinetic energy of the flowing water to impulse a turbomachine, and in turn the turbomachine turns a generator to generate electricity, this process is reversible because the system and environment could be recovered in such a way that the water is carried back to the reservoir with a pump driven by an electromotor with the electrical energy.

2.6.2 Energy Equation in Differential Form for a Steady Flow

For a steady flow the equation (2-38) might be written in differential form

$$\delta q_H - \delta w_s = d\left(\frac{p}{\rho}\right) + g dz + u du + d e_i \tag{2-72}$$

from thermodynamics

$$T ds = d e_i + p d\upsilon \tag{2-73}$$

in which s is the entropy per unit mass of fluid and υ is the specific volume of fluid, $\upsilon = 1/\rho$.

By substituting the equation (2-73) into the equation (2-72) and noticing, $d(p/\rho) = \upsilon dp + p d\upsilon = dp/\rho + p d\upsilon$

$$\delta q_H - \delta w_s - T ds = \frac{dp}{\rho} + g dz + u du \tag{2-74}$$

2.6.3 Energy Equation for a Reversible Process

For a reversible process, $\delta q_H = T ds$, if there is no work done by the system on its surroundings or by the surroundings to the system, $\delta w_s = 0$, the equation (2-74) reduces to

$$\frac{dp}{\rho} + g dz + u du = 0 \tag{2-75}$$

which is the energy equation in differential form for the reversible process and it is the same with the Euler's equation of the equation (2-64). The equation (2-75) also hold true for the reversible process in an adiabatic flow, in which $\delta q_H = 0$, $ds = 0$.

2.6.4 Energy Equation for an Irreversible Process

For an irreversible flow the relationship of the entropy and the heat in system is the Clausius inequality[①]

$$T ds - \delta q_H > 0 \tag{2-76}$$

Let the flow losses per unit weight of fluid be h_w.

$$g h_w = T ds - \delta q_H \tag{2-77}$$

therefore, the equation (2-74) is rewritten as

$$\delta w_s + \frac{dp}{\rho} + g dz + u du + g h_w = 0 \tag{2-78}$$

which is the energy equation in differential form for an irreversible process.

If $\delta w_s = 0$

$$\frac{dp}{\rho} + g dz + u du + g h_w = 0 \tag{2-79}$$

For incompressible fluid in the adiabatic flow, ρ = constant, $de_i = c_V dT$ and $\delta q_H = 0$. If $\delta w_s = 0$, the equation (2-72) becomes

$$\frac{dp}{\rho} + g dz + u du + c_V dT = 0 \tag{2-80}$$

Comparing the equation (2-80) with the equation (2-79), it can be seen that the losses in mechanical energy make the increase in temperature of fluid in an irreversible process.

2.6.5 Bernoulli's Equation for Real System

If there is work done by the fluid system on its surroundings or by the surroundings to the fluid system, the Bernoulli's equation in a real system can be obtained by the equation (2-78), and it is

$$\frac{u_1^2}{2g} + \frac{p_1}{\rho g} + z_1 \pm h_s = \frac{u_2^2}{2g} + \frac{p_2}{\rho g} + z_2 + h_w \tag{2-81}$$

in which h_s is the shaft work per unit weight of fluid, $h_s = w_s/g$, and the sign: " + " denotes the energy added to the system, " − " denotes the energy gotten away the system to the surrounding. h_w is the energy losses per unit weight of fluid from section 1-1 to section 2-2 and is always positive, which shows that the total energy decreases gradually in the downstream direction.

2.7 Application of the Energy Equation to Steady Fluid-Flow Situations

The equation (2-72) can be only applied to a ministream tube or a streamline because on the section of ministream tube the physical parameters (such as elevation z, pressure p and velocity u) of each fluid particle could be considered as the same. In the engineering practice, it is not enough to know what the energy equation for a streamline is, and the most important is what the energy equation for a total flow would be.

2.7.1 Gradually Varied Flow

In convenience for the calculation of parameters on a cross section, the concept of gradually varied flow is discussed before the Bernoulli's equation for a total flow is established.

The gradually varied flow is defined as that on the flow section the angle β between two streamlines is very small or the radius of curvature R of each streamline is infinity as shown in Figure 2.13. All the flows which do not satisfy the qualifications are called the rapidly varied flow. The flows at section $a-a$ and section $b-b$ in Figure 2.3 are the gradually varied flow. The flow at section $1-1$ in Figure 2.3 is rapidly varied flow. The essence of gradually varied flow is that: the streamlines tend to be in the parallel beelines; the acceleration is very small in the flow and the inertia force could be neglected; the flow section can be considered a plane surface and for this reason the flow section is commonly taken as the cross section.

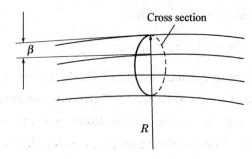

Figure 2.13 Cross section with gradually varied flow

In the flow section $n-n$ of gradually varied flow, an element fluid column with area $\mathrm{d}A$ and length $\mathrm{d}z$ is selected as the free body as shown in Figure 2.14.

Chapter 2 Foundation of Fluid Mechanics

The upper surface of column is coincident with the streamline ab and the lower surface with the streamline cd. The elevation of the lower surface of column is the coordinate z. The pressure on the lower surface is p and the pressure on the upper surface is $p + \mathrm{d}p$. The velocity of fluid column is u. There are forces acting on the column in the $n - n$ direction not only the pressure force, but also the forces including: the weight of element fluid, $\rho g \mathrm{d}A \mathrm{d}z$; the centrifugal force of element fluid $(\rho g \mathrm{d}A \mathrm{d}z / g)(u^2/R)$. According to the conditions of gradually varied flow, the curvature radius is infinite, so the centrifugal force is very small, it could be neglected; the surface forces of element fluid including the internal frictional force on the two end surfaces and pressure forces on the circumferential surface of column is orthogonal with the axis $n - n$, so there are no components of them in the n direction. For steady flow, $\partial u / \partial t = 0$, the algebraic sum of all forces is zero in the n direction, namely

$$(p + \mathrm{d}p)\mathrm{d}A - p\mathrm{d}A + \rho g \mathrm{d}A \mathrm{d}z = 0 \quad (2-82)$$

After dividing through by $\rho g \mathrm{d}A$, the equation reduces to

$$\frac{\mathrm{d}p}{\rho g} + \mathrm{d}z = 0 \quad (2-83)$$

For incompressible fluid, ρ is a constant, so

$$d\left(\frac{p}{\rho g} + z\right) = 0 \quad (2-84)$$

by integrating the equation reduces to

$$\frac{p}{\rho g} + z = C \quad (2-85)$$

The equation says that the distribution of pressure at each point on various streamlines over the cross section of gradually varied flow is the same as the distribution of pressure for the statics. In other words, $p/(\rho g) + z$ is a constant over the whole cross section, but the constant is different for the various sections.

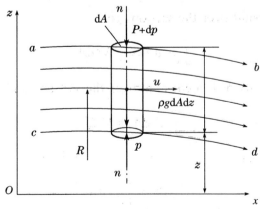

Figure 2.14 Control volume in a flow section with gradually varied flow

2.7.2 Bernoulli's Equation for the Real-Fluid Total Flow

The total flow consists of countless ministream tube, the mechanical energy per unit weight of fluid particle in a ministream tube is

$$e = \frac{u^2}{2g} + \frac{p}{\rho g} + z \tag{2-86}$$

When fluid with weight $dG = \rho g u dA$ pass through the ministream tube, the energy of fluid over the cross section is $dE = e dG$. Then, the total energy per unit time through the cross section is

$$E = \int_A dE = \int_A \left(\frac{u^2}{2g} + \frac{p}{\rho g} + z \right) \rho g u dA \tag{2-87}$$

If the conditions of gradually varied flow on the section are satisfied, $p/(\rho g) + z$ could be taken out from the integral sign, the above equation becomes

$$E = \left(\frac{p}{\rho g} + z \right) \rho g \int_A u dA + \frac{\rho}{2} \int_A u^3 dA \tag{2-88}$$

Because of $\int_A u dA = Q$, the first term in the equation is

$$\left(\frac{p}{\rho g} + z \right) \rho g Q \tag{2-89}$$

The distribution of velocity over the cross section is not uniform, so an average velocity V is employed

$$V = \frac{Q}{V} = \frac{\int_A u dA}{A} \tag{2-90}$$

The velocity u is the velocity at a point and it is not the same as the average velocity V in every point over the section, so

$$\int_A u^3 dA \neq \int_A V^3 dA \tag{2-91}$$

By noticing velocity V is a constant over the section, and

$$\int_A V^3 dA = V^2 Q \tag{2-92}$$

For this reason, let

$$\alpha V^3 = \int_A u^3 dA \tag{2-93}$$

then

Chapter 2　Foundation of Fluid Mechanics

$$\frac{\rho}{2}\int_A u^3 \mathrm{d}A = \frac{\alpha V^2}{2g}\rho g Q \tag{2-94}$$

The two integrals of the equation (2-88) have been obtained, the energy over the section in the gradually varied flow is

$$E = \left(\frac{p}{\rho g} + z\right)\rho g Q + \frac{\alpha V^2}{2g}\rho g Q \tag{2-95}$$

in which α is the kinetic-energy[①] correction factor which is the result of the nonuniform distribution of velocity over the section. For turbulent flow in a pipe, $\alpha = 1.05 \sim 1.10$. For laminar flow in a pipe, $\alpha = 2$.

Because the extent of turbulence is intensive in mostly practical engineering, it is usually $\alpha = 1$ for the ordinarily calculation except for precise work.

When each term of above formula is divided by the weight flow rate, $\rho g Q$ passing through the cross section, the mechanical energy per unit weight over the section in gradually varied flow is

$$e = \left(\frac{p}{\rho g} + z\right) + \frac{\alpha V^2}{2g} \tag{2-96}$$

Let h_w be the energy losses per unit weight of fluid from section 1-1 to section 2-2, the Bernoulli's equation for a total flow is

$$\frac{\alpha_1 V_1^2}{2g} + \frac{p_1}{\rho g} + z_1 = \frac{\alpha_2 V_2^2}{2g} + \frac{p_2}{\rho g} + z_2 + h_w \tag{2-97}$$

Let h_s be the shaft work per unit weight of fluid, the Bernoulli's equation is

$$\frac{\alpha_1 V_1^2}{2g} + \frac{p_1}{\rho g} + z_1 \pm h_s = \frac{\alpha_2 V_2^2}{2g} + \frac{p_2}{\rho g} + z_2 + h_w \tag{2-98}$$

2.7.3　Applications of Bernoulli's Equation

Bernoulli's energy equation is one of the basic equations in the hydrodynamics. It plays an important role in solution of engineering problems. The restrictions which should be obeyed in applications are summed up as follows:

(1) The fluid is incompressible and the flow must be a steady flow.

(2) The selected cross section should be in the gradually varied flow or in uniform flow. But the flow between the two sections could be both the gradually varied flow and the rapidly varied flow. The definition of the gradually varied flow is uncertain in the measure of angle and radius, an angle $\beta < 5°$ between two streamlines is given here for reference. In practical applications, the sections selected are usually such as the free surface in a big container, the vena contracta in an orifice, the transverse section in a long straight tube with constant diameter, and

so on.

(3) If there is input or output of shaft work between two sections, the calculation can be carried out with the equation (2-98).

(4) The flow rate remains a constant along the flow path. In a flow with branch or confluence the energy equation can be given in according with the conservation law of energy. In Figure 2.15, the flow Q_1 in section 1-1 is divided into two branches of flow Q_2 and Q_3 from the continuity equation, $Q_1 = Q_2 + Q_3$, so, for these three sections the energy equation can be expressed as follows

$$\left(\frac{\alpha_1 V_1^2}{2g} + \frac{p_1}{\rho g} + z_1\right)Q_1 = \left(\frac{\alpha_2 V_2^2}{2g} + \frac{p_2}{\rho g} + z_2\right)Q_2 + \left(\frac{\alpha_3 V_3^2}{2g} + \frac{p_3}{\rho g} + z_3\right)Q_3 +$$
$$h_{wl-2}Q_2 + h_{wl-3}Q_3 \qquad (2-99)$$

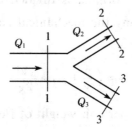

Figure 2.15 Branching flow

(5) For a velocity below 50 m/s, the gas could be treated approximately as the incompressible fluid. If the velocity of gas is over 50 m/s, the Bernoulli's equation for compressible fluid should be used for a high accuracy.

2.8 Applications of the Linear-Momentum Equation

The linear-momentum equation was obtained by Reynolds transport equation (2-32).

This equation may be written in the component form, say the x direction

$$\sum F_x = \frac{\partial}{\partial t}\iiint_\tau \rho u_x d\tau + \iint_A (\rho u_x) \boldsymbol{u} \cdot d\boldsymbol{A} \qquad (2-100)$$

In words, for any direction the sum of variation of the momentum within a control volume plus the net outflow of momentum from the control volume is equal to the component of the resultant external force acting on a control volume at that direction in unit time.

For a steady flow of incompressible fluid, if the control surfaces, normal to the velocity wherever it cuts across the flow, have only the two surfaces, i.e., the outflow surface and the inflow surface, and if the velocities in the surfaces are uniform, the surface integral in the equation (2-100) could be removed. It reduces to

$$\sum F = \rho u_2 u_2 A_2 - \rho u_1 u_1 A_1 \tag{2-101}$$

By making use of the continuity equation (2-52), the above equation reduces to

$$\sum F = \rho Q(u_2 - u_1) \tag{2-102}$$

In the actual flow of incompressible fluid the velocity over a plane cross section is not uniform. A dimensionless momentum correction factor β is introduced. Let V be the average velocity over the section.

$$\int_A \rho u^2 dA = \beta \rho V^2 A \tag{2-103}$$

or

$$\beta = \frac{1}{A} \int_A \left(\frac{u}{V}\right)^2 dA \tag{2-104}$$

The momentum correction factor β is a coefficient not less than 1. In a straight round tube, it is 4/3 for laminar flow. β has a experimental value of 1.02~1.05 in turbulent flow and it could be taken as 1.

If the un-uniform in velocity over the section is taken into account, the equation (2-102) may be rewritten as

$$\sum F = \rho Q(\beta_2 V_2 - \beta_1 V_1) \tag{2-105}$$

Notes

① time-varying acceleration 时变加速度
② reducing pipe 减径管，渐缩管
③ winding pipe 弯管，风管
④ no-flow control surfaces 非通流控制面
⑤ Lagrange mean value theorem 拉格朗日中值定理
⑥ dash-dotted line 点划线
⑦ Clausius inequality 克劳修斯不等式
⑧ kinetic-energy 动能的

Vocabulary

adiabatic [ˌeɪdaɪəˈbætɪk] adj. [物]绝热的；隔热的
algebraic [ˌældʒɪˈbreɪɪk] adj. [数]代数的；关于代数学的
beeline [ˈbiːlaɪn] n. 直线
contracta [kənˈtræktə] n. 缩脉；流颈
convective [kənˈvektɪv] adj. 对流的；传递性的
curvature [ˈkɜːvətʃə] n. 弯曲；[数]曲率
curvous [ˈkɜːvəs] adj. 弯曲的

Chapter 2 Foundation of Fluid Mechanics

cylinder ['sɪlɪndə] n. 圆筒;汽缸;[数]柱面;圆柱状物

datum ['deɪtəm] n. 数据,资料;[测]基点,基线,基面;论据,作为论据的事实

derivative [dɪ'rɪvətɪv] n. [化学]衍生物,派生物;导数 adj. 派生的;引出的

differential [ˌdɪfə'renʃ(ə)l] n. 微分;差别 adj. 微分的;差别的;特异的

dimensionality [dɪˌmenʃə'nælətɪ] n. 维数,维度,量纲

drainage ['dreɪnɪdʒ] n. 排水;排水系统;污水;排水面积

dynamics [daɪ'næmɪks] n. 动力学,力学

equivalency [ɪ'kwɪvələnsɪ] n. 等价;相等(等于 equivalence)

hexahedron [ˌheksə'hiːdrən] n. [数]六面体

ineluctability [ˌɪnɪlʌktə'bɪlətɪ] n. 不可避免,无法逃避

inequality [ˌɪnɪ'kwɒlɪtɪ] n. 不平等;不同;不平均

inflow ['ɪnfləʊ] n. 流入;流入物;岭量 vi. 流入;货币回笼

irreversible [ˌɪrɪ'vɜːsɪb(ə)l] adj. 不可逆的;不能取消的;不能翻转的

irrotational [ˌɪrəʊ'teɪʃ(ə)n(ə)l] adj. 无旋的;无漩涡的

laminar ['læmɪnə] adj. 层状的,薄片状的;板状的

momentum [mə'mentəm] n. 势头;[物]动量;动力

orifice ['ɒrɪfɪs] n. [机]孔口

outflow ['aʊtfləʊ] n. 流出;流出量;流出物 vi. 流出

parabola [pə'ræb(ə)lə] n. 抛物线

pervade [pə'veɪd] vt. 遍及;弥漫

piezometer [ˌpaɪɪ'zɒmɪtə] n. [物]压强计

reversible [rɪ'vɜːsɪb(ə)l] adj. 可逆的;可撤消的;可反转的

rotational [rəʊ'teɪʃənəl] adj. 转动的;回转的;轮流的

tangent ['tændʒənt] n. [数]切线,[数]正切 adj. 切线的,相切的

temporal ['temp(ə)r(ə)l] n. 世间万物;暂存的事物 adj. 暂时的;当时的;现世的

trajectory [trə'dʒekt(ə)rɪ] n. [物]轨道,轨线;[航][军]弹道

turbulent ['tɜːbjʊl(ə)nt] adj. 激流的,湍流的

vena ['viːnə] n. 脉;[解剖]静脉

viscosity [vɪ'skɒsɪtɪ] n. [物]粘性,[物]粘度

volumetric [ˌvɒljʊ'metrɪk] adj. [物]体积的;[物]容积的;[物]测定体积的

vortex ['vɔːteks] n. [航][流]涡流;漩涡;(动乱,争论等的)中心;旋风

◆ Exercises

1. Translate the following Chinese phrases into English

(1) 空间位置点　　(2) 粘性力　　(3) 迁移加速度

(4) 层流　　(5) 时均速度　　(6) 边界条件

(7) 通流面　　(8) 通流控制面　　(9) 实线

(10) 可逆过程　　(11) 速度分布　　(12) 实际工程

(13) 横断面　　(14) 控制体积　　(15) 轨迹线

(16) 内点　　　　　　　　(17) 基准线

2. Translate the following English phrases into Chinese

(1) fluid particles　　　　　　　　(2) substantial derivative
(3) turbulent flow　　　　　　　　(4) unsteady flow
(5) uniform flow　　　　　　　　(6) total flow
(7) volume flow rate　　　　　　　(8) continuity equation
(9) differential form　　　　　　　(10) physical parameters
(11) incompressible fluid　　　　　(12) vena contracta
(13) fluid mechanics　　　　　　　(14) general property
(15) vortex flow　　　　　　　　　(16) net mass flux

3. Translate the following Chinese sentences into English

(1) 这就是雷诺输运定理,它表示系统内 N 的变化率等于控制体内 N 的变化率加上 N 流出控制体的净流量。

(2) 稳定流动的意思是说流体流动状态不随时间而变化。

(3) 由于流线上流体质点的速度总是与流线相切,垂直于流线的速度分量为零,所以流体不能穿过流管流入或流出。

(4) 伯努利方程有非常广泛的应用。因此,这里只介绍几个常见的应用伯努利方程的实例。

(5) 迹线是在流场空间所作的一条曲线,其确定了给定的流体质点所经过的轨迹。

4. Translate the following English sentences into Chinese

(1) It is the rate of change with respect to time of the total amount of the physical property inside the control volume at time t.

(2) The dynamics of rigid bodies is a fascinating and complicated subject.

(3) A fluid is a continuum consisting of numerous fluid particles. The space filled with flowing fluids is called the flow field.

(4) Stream tube with infinitesimal cross-sectional area is called elementary stream tube, and the limitation of it is streamline.

(5) According to how many space variables flow parameters are depended upon, fluid flow may be classified as one-dimensional, two-dimensional and three-dimensional flow.

(6) Generally, there are two approaches used to describe the motion of a fluid in the study of fluid mechanics, namely, they are Lagrangian approach and Eulerian approach.

5. Translate the following Chinese essay into English

流体力学是研究流体平衡与宏观机械运动规律的一门学科,是力学最重要的分支之一。流体力学按其研究内容侧重方面的不同,分为理论流体力学(通称流体力学)和应用流体力学(通称工程流体力学)。前者主要采用严密的数学推理方法,力求准确性

Chapter 2 Foundation of Fluid Mechanics

和严密性。后者则侧重于解决工程实际中出现的问题,而不去追求数学上的严密性。

6. Translate the following English essay into Chinese

The continuity equation is basedupon the conservation law of mass as it applies to the flow of fluids. In other words, the continuity equation manifests that the mass rate of outflow from a certain region of space minus that of inflow into the same region, is equal to mass increment rate of the region. There are types of continuity equation, namely, continuity equation in integral form and differential form.

扫一扫,查看更多资料

Chapter 3 Foundation of Heat Transfer

3.1 Introduction

Heat transfer is the science that seeks to predict the energy transfer that may take place between material bodies as a result of a temperature difference. Thermodynamics teaches that this energy transfer is defined as heat. The science of heat transfer seeks not merely to explain how heat energy may be transferred, but also to predict the rate at which the exchange will take place under certain specified conditions. The fact that a heat-transfer rate is the desired objective of an analysis points out the difference between heat transfer and thermodynamics. Thermodynamics deals with systems in equilibrium; it may be used to predict the amount of energy required to change a system from one equilibrium state to another; it may not be used to predict how fast a change will take place since the system is not in equilibrium during the process. Heat transfer supplements the first and second principles of thermodynamics by providing additional experimental rules that may be used to establish energy-transfer rates. As in the science of thermodynamics, the experimental rules used as a basis of the subject of heat transfer are rather simple and easily expanded to encompass a variety of practical situations.

As an example of the different kinds of problems treated by thermodynamics and heat transfer, consider the cooling of a hot steel bar that is placed in a pail of water. Thermodynamics may be used to predict the final equilibrium temperature of the steel bar-water combination. Thermodynamics will not tell us how long it takes to reach this equilibrium condition or what the temperature of the bar will be after a certain length of time before the equilibrium condition is attained. Heat transfer may be used to predict the temperature of both the bar and the water as a function of time.

Most readers will be familiar with the terms used to denote the three modes of

Chapter 3 Foundation of Heat Transfer

heat transfer: conduction, convection and radiation. In this chapter we seek to explain the mechanism of these modes qualitatively so that each may be considered in its proper perspective. Subsequent chapters treat the three types of heat transfer in detail.

3.2 Conduction Heat Transfer

When a temperature gradient exists in a body, experience has shown that there is an energy transfer from the high-temperature region to the low-temperature region. We say that the energy is transferred by conduction and that the heat-transfer rate per unit area is proportional to the normal temperature gradient:

$$\frac{q_x}{A} \sim \frac{\partial T}{\partial x} \tag{3-1}$$

When the proportionality constant is inserted,

$$q_x = -kA\frac{\partial T}{\partial x} \tag{3-2}$$

where q_x is the heat-transfer rate and $\partial T/\partial x$ is the temperature gradient in the direction of the heat flow. The positive constant k is called the thermal conductivity of the material, and the minus sign is inserted so that the second principle of thermodynamics will be satisfied; i.e., heat must flow downhill on the temperature scale, as indicated in the coordinate system of Figure 3.1. The equation (3-2) is called Fourier's law[①] of heat conduction after the French mathematical physicist Joseph Fourier, who made very significant contributions to the analytical treatment of conduction heat transfer. It is important to note that the equation (3-2) is the defining equation for the thermal conductivity and that k has the units of watts per meter per Celsius degree in a typical system of units in which the heat flow is expressed in watts.

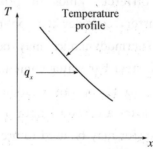

Figure 3.1 Sketch showing direction of heat flow

We now set ourselves the problem of determining the basic equation that governs the transfer of heat in a solid, using the equation (3-2) as a starting point.

Consider the one-dimensional system shown in Figure 3.2. If the system is in a steady state, i.e., if the temperature does not change with time, then the problem is a simple one, and we need only integrate the equation (3-2) and substitute the appropriate values to solve for the desired quantity. However, if the temperature of the solid is changing with time; or if there are heat sources or sinks within the solid, the situation is more complex. We consider the general case where the temperature may be changing with time and heat sources may be present within the body. For the element of thickness dx, the following energy balance may be made:

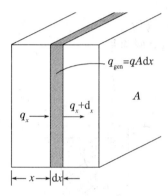

Figure 3.2 Elemental volume for one-dimension heat-conduction analysis

Energy conducted in left face + heat generated within element
= change in internal energy + energy conducted out right face

These energy quantities are given as follows:

$$\text{Energy in left face} = q_x = -kA \frac{\partial T}{\partial x} \tag{3-3}$$

$$\text{Energy generated within element} = \dot{q} A dx \tag{3-4}$$

$$\text{Change in internal energy} = \rho c A \frac{\partial T}{\partial \tau} dx \tag{3-5}$$

$$\text{Energy out right face} = q_x + dx = -kA \frac{\partial T}{\partial x} \bigg]_{x+dx} \tag{3-6}$$

$$= -A \left[k \frac{\partial T}{\partial x} + \frac{\partial}{\partial x} \left(k \frac{\partial T}{\partial x} \right) dx \right]$$

where
$\dot{q}$ = energy generated per unit volume, W/m³.
c = specific heat of material, J/kg·℃.
ρ = density, kg/m³.

Combining the relations above gives.

$$-kA\frac{\partial T}{\partial x} + \dot{q}A\mathrm{d}x = \rho c A\frac{\partial T}{\partial \tau}\mathrm{d}x - A\left[k\frac{\partial T}{\partial x} + \frac{\partial}{\partial x}\left(k\frac{\partial T}{\partial x}\right)\mathrm{d}x\right] \quad (3-7)$$

or

$$\frac{\partial}{\partial x}\left(k\frac{\partial T}{\partial x}\right) + \dot{q} = \rho c\frac{\partial T}{\partial \tau} \quad (3-8)$$

This is the one-dimensional heat-conduction equation. To treat more than one-dimensional heat flow, we need consider only the heat conducted in and out of a unit volume in all three coordinate directions, as shown in Figure 3.3a.

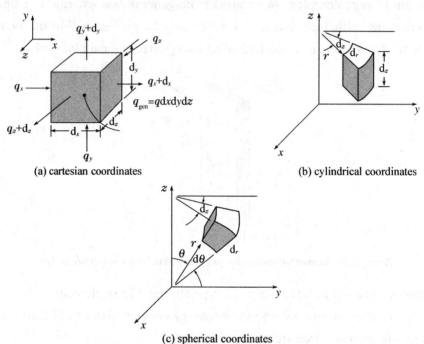

(a) cartesian coordinates (b) cylindrical coordinates

(c) spherical coordinates

Figure 3.3 Elemental volume for three-dimensional heat-conduction analysis

The energy balance yields

$$q_x + q_y + q_z + q_{gen} = q_{x+\mathrm{d}x} + q_{y+\mathrm{d}y} + q_{z+\mathrm{d}z} + \frac{\mathrm{d}E}{\mathrm{d}\tau} \quad (3-9)$$

and the energy quantities are given by

$$q_x = -k\mathrm{d}y\mathrm{d}z\frac{\partial T}{\partial x}$$

$$q_{x+\mathrm{d}x} = -\left[k\frac{\partial T}{\partial x} + \frac{\partial}{\partial x}\left(k\frac{\partial T}{\partial x}\right)\mathrm{d}x\right]\mathrm{d}y\mathrm{d}z$$

$$q_y = -k\mathrm{d}x\mathrm{d}z\frac{\partial T}{\partial y}$$

$$q_{y+\mathrm{d}y} = -\left[k\frac{\partial T}{\partial y} + \frac{\partial}{\partial y}\left(k\frac{\partial T}{\partial y}\right)\mathrm{d}y\right]\mathrm{d}x\mathrm{d}z \quad (3-10)$$

$$q_z = -k\mathrm{d}x\mathrm{d}y\frac{\partial T}{\partial z}$$

$$q_{z+dz} = -\left[k\frac{\partial T}{\partial z} + \frac{\partial}{\partial z}\left(k\frac{\partial T}{\partial z}\right)dz\right]dxdy$$

$$q_{gen} = -\dot{q}dxdydz$$

$$\frac{dE}{d\tau} = \rho c\, dxdydz\frac{\partial T}{\partial \tau}$$

so that the general three-dimensional heat-conduction equation is

$$\frac{\partial}{\partial x}\left(k\frac{\partial T}{\partial x}\right) + \frac{\partial}{\partial y}\left(k\frac{\partial T}{\partial y}\right) + \frac{\partial}{\partial z}\left(k\frac{\partial T}{\partial z}\right) + \dot{q} = \rho c\frac{\partial T}{\partial \tau} \qquad (3-11)$$

For constant thermal conductivity, the equation (3-11) is written

$$\frac{\partial^2 T}{\partial x^2} + \frac{\partial^2 T}{\partial y^2} + \frac{\partial^2 T}{\partial z^2} + \frac{\dot{q}}{k} = \frac{1}{\alpha}\frac{\partial T}{\partial \tau} \qquad (3-12)$$

where the quantity $\alpha = k/\rho c$ is called the thermal diffusivity of the material. The larger the value of α, the faster heat will diffuse through the material. This may be seen by examining the quantities that make up α. A high value of α could result either from a high value of thermal conductivity, which would indicate a rapid energy-transfer rate, or from a low value of the thermal heat capacity ρc. A low value of the heat capacity would mean that less of the energy moving through the material would be absorbed and used to raise the temperature of the material; thus more energy would be available for further transfer. Thermal diffusivity α has units of square meters per second.

In the derivations above, the expression for the derivative at $x + dx$ has been written in the form of a Taylor-series expansion with only the first two terms of the series employed for the development.

The equation (3-12) may be transformed into either cylindrical or spherical coordinates by standard calculus techniques. The results are as follows.

Cylindrical coordinates

$$\frac{\partial^2 T}{\partial r^2} + \frac{1}{r}\frac{\partial T}{\partial r} + \frac{1}{r^2}\frac{\partial^2 T}{\partial \varphi^2} + \frac{\partial^2 T}{\partial z^2} + \frac{\dot{q}}{k} = \frac{1}{\alpha}\frac{\partial T}{\partial \tau} \qquad (3-13)$$

Spherical coordinates

$$\frac{1}{r}\frac{\partial^2}{\partial r^2}(rT) + \frac{1}{r^2\sin\theta}\frac{\partial}{\partial \theta}\left(\sin\theta\frac{\partial T}{\partial \theta}\right) + \frac{1}{r^2\sin^2\theta}\left(\sin\theta\frac{\partial^2 T}{\partial \varphi^2}\right) + \frac{\dot{q}}{k} = \frac{1}{\alpha}\frac{\partial T}{\partial \tau}$$

$$(3-14)$$

The coordinate systems for use with the equations (3-13) and (3-14) are indicated in Figure 3.3b and Figure 3.3c, respectively.

Many practical problems involve only special cases of the general equations listed above. As a guide to the developments in future chapters, it is worthwhile to show the reduced form of the general equations for several cases of practical interest.

Steady-state one-dimensional heat flow (no heat generation)

$$\frac{d^2 T}{dx^2} = 0 \tag{3-15}$$

Note that this equation (3-15) is the same as the equation (3-2) when q = constant.

Steady-state one-dimensional heat flow in cylindrical coordinates (no heat generation)

$$\frac{d^2 T}{dr^2} + \frac{1}{r}\frac{dT}{dr} = 0 \tag{3-16}$$

Steady-state one-dimensional heat flow with heat sources

$$\frac{d^2 T}{dr^2} + \frac{\dot{q}}{k} = 0 \tag{3-17}$$

Two-dimensional steady-state conduction without heat sources

$$\frac{\partial^2 T}{\partial x^2} + \frac{\partial^2 T}{\partial y^2} = 0 \tag{3-18}$$

3.3 Thermal Conductivity

The equation (3-2) is the defining equation for thermal conductivity. On the basis of this definition, experimental measurements may be made to determine the thermal conductivity of different materials. For gases at moderately low temperatures, analytical treatments in the kinetic theory of gases may be used to predict accurately the experimentally observed values. In some cases, theories are available for the prediction of thermal conductivities in liquids and solids, but in general, many open questions and concepts still need clarification where liquids and solids are concerned.

The mechanism of thermal conduction in a gas is a simple one. We identify the kinetic energy of a molecule with its temperature; thus, in a high-temperature region, the molecules have higher velocities than in some lower-temperature region. The molecules are in continuous random motion, colliding with one another and exchanging energy and momentum. The molecules have this random motion whether or not a temperature gradient exists in the gas. If a molecule moves from a high-temperature region to a region of lower temperature, it transports kinetic energy to the lower-temperature part of the system and gives up this energy through collisions with lower-energy molecules.

Table 3.1 lists typical values of the thermal conductivities for several materials to indicate the relative orders of magnitude to be expected in practice.

Chapter 3 Foundation of Heat Transfer

Table 3.1 Thermal conductivity of various materials at 0 ℃

Material	Thermal conductivity	
	W / m · ℃	Btu / h · ft · °F
Metals		
Silver (pure)	410	237
Copper (pure)	385	223
Aluminum (pure)	202	117
Nickel (pure)	93	54
Iron (pure)	73	42
Carbon steel, 1 % C	43	25
Lead (pure)	35	20.3
Chrome-nickel steel (18 % Cr, 8 % Ni)	16.3	9.4
Nonmetallic solids		
Diamond	2 300	1 329
Quartz, parallel to axis	41.6	24
Magnesite	4.15	2.4
Marble	2.08~2.94	1.2~1.7
Sandstone	1.83	1.06
Glass, window	0.78	0.45
Maple or oak	0.17	0.096
Hard rubber	0.15	0.087
Polyvinyl chloride	0.09	0.052
Styrofoam	0.033	0.019
Sawdust	0.059	0.034
Glass wool	0.038	0.022
Ice	2.22	1.28
Liquids		
Mercury	8.21	4.74
Water	0.556	0.327
Ammonia	0.540	0.312
Lubricating oil, SAE 50	0.147	0.085
Freon 12, CCl_2F_2	0.073	0.042
Gases		
Hydrogen	0.175	0.101
Helium	0.141	0.081
Air	0.024	0.013 9
Water vapor (saturated)	0.020 6	0.011 9
Carbon dioxide	0.014 6	0.008 44

We noted that thermal conductivity has the units of watts per meter per Celsius degree when the heat flow is expressed in watts. Note that a heat rate is involved, and the numerical value of the thermal conductivity indicates how fast heat will flow in a given material. How is the rate of energy transfer taken into account in the molecular model discussed above? Clearly, the faster the molecules move, the faster they will transport energy. Therefore, the thermal conductivity of a gas should be dependent on temperature. A simplified analytical treatment shows the thermal conductivity of a gas to vary with the square root of the absolute temperature. (It may be recalled that the velocity of sound in a gas varies with the square root of the absolute temperature; this velocity is approximately the mean speed of the molecules.) Thermal conductivities of some typical gases are shown in Figure 3.4. For most gases at moderate pressures the thermal conductivity is a function of temperature alone. This means that the gaseous data for 1 atmosphere (atm). When the pressure of the gas becomes of the order of its critical pressure or, more generally, when nonideal-gas behavior is encountered, other sources must be consulted for thermal-conductivity data.

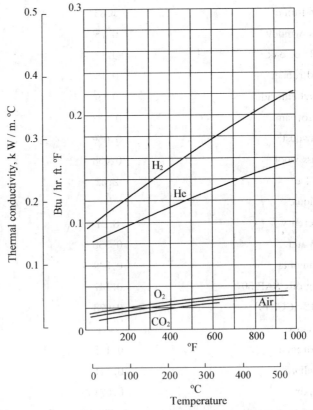

Figure 3.4 **Thermal conductivities of some typical gases**

$1 \text{ W/m} \cdot \text{°C} = 0.5779 \text{ Btu/h} \cdot \text{ft} \cdot \text{°F}$

The physical mechanism of thermal-energy conduction in liquids is qualitatively the same as in gases; however, the situation is considerably more complex because the molecules are more closely spaced and molecular force fields exert a strong influence on the energy exchange in the collision process. Thermal conductivities of some typical liquids are shown in Figure 3.5.

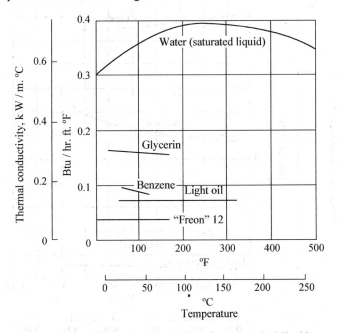

Figure 3.5 Thermal conductivities of some typical liquids

In the English system of units, heat flow is expressed in British thermal units per hour (Btu / h), area in square feet, and temperature in degrees Fahrenheit. Thermal conductivity will then have units of Btu / h · ft · °F.

Thermal energy may be conducted in solids by two modes: lattice vibration and transport by free electrons. In good electrical conductors a rather large number of free electrons move about in the lattice structure of the material. Just as these electrons may transport electric charge, they may also carry thermal energy from a high-temperature region to a low-temperature region, as in the case of gases. In fact, these electrons are frequently referred to as the electron gas. Energy may also be transmitted as vibrational energy in the lattice structure of the material. In general, however, this latter mode of energy transfer is not as large as the electron transport, and for this reason good electrical conductors are almost always good heat conductors, namely, copper, aluminum and silver, and electrical insulators are usually good heat insulators. A notable exception is diamond, which is an electrical insulator, but which can have a thermal conductivity five times as high as silver or copper. It is this fact that enables a jeweler to distinguish between genuine diamonds and fake stones. A small instrument is available that measures the response of the stones to a thermal heat pulse. A true diamond will exhibit a far more rapid response than the nongenuine stone.

Thermal conductivities of some typical solids are shown in Figure 3.6.

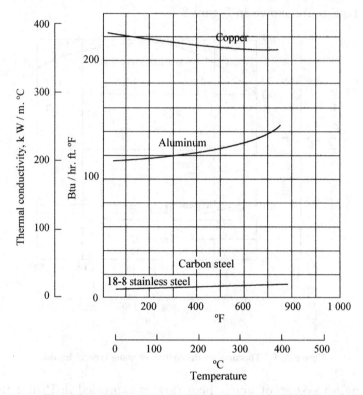

Figure 3.6 Thermal conductivities of some typical solids

An important technical problem is the storage and transport of cryogenic liquids like liquid hydrogen over extended periods of time. Such applications have led to the development of super insulations for use at these very low temperatures (down to about $-250\ ℃$). The most effective of these super insulations consists of multiple layers of highly reflective materials separated by insulating spacers. The entire system is evacuated to minimize air conduction and thermal conductivities as low as $0.3\ W/m \cdot ℃$ are possible. A convenient summary of the thermal conductivities of a few insulating materials at cryogenic temperatures is given in Table 3.2.

Table 3.2 Effective thermal conductivities of cryogenic insulating materials for use in range 15 ℃ to −195 ℃ and density range 30 to 80 kg/m³

Type of insulation	Effective k, mW/m · ℃
Foams, powders, and fibers, unevacuated	7~36
Powders, evacuated	0.9~6

	(to be continued)
Type of insulation	Effective k, mW/m · ℃
Glass fibers, evacuated	0.6~3
Opacified powders, evacuated	0.3~1
Multilayer insulations, evacuated	0.015~0.06

3.4 Convection Heat Transfer

It is well known that a hot plate of metal will cool faster when placed in front of a fan than when exposed to still air. We say that the heat is convected away, and we call the process convection heat transfer. The term convection, provides the reader with an intuitive notion concerning the heat-transfer process; however, this intuitive notion must be expanded to enable one to arrive at anything like an adequate analytical treatment of the problem. For example, we know that the velocity at which the air blows over the hot plate obviously influences the heat-transfer rate. But does it influence the cooling in a linear way; i.e., if the velocity is doubled, will the heat-transfer rate double? We should suspect that the heat-transfer rate might be different if we cooled the plate with water instead of air, but, again, how much difference would there be? These questions may be answered with the aid of some rather basic analyses. For now, we sketch the physical mechanism of convection heat transfer and show its relation to the conduction process.

Consider the heated plate shown in Figure 3.7. The temperature of the plate is T_w, and the temperature of the fluid is T_∞. The velocity of the flow will appear as shown, being reduced to zero at the plate as a result of viscous action. Since the velocity of the fluid layer at the wall will be zero, the heat must be transferred only by conduction at that point. Thus, we might compute the heat transfer, using the equation (3 - 2), with the thermal conductivity of the fluid and the fluid temperature gradient at the wall. Why, then, if the heat flows by conduction in this layer, do we speak of convection heat transfer and need to consider the velocity of the fluid? The answer is that the temperature gradient is dependent on the rate at, which the fluid carries the heat away; a high velocity produces a large temperature gradient, and so on. Thus, the temperature gradient at the wall depends on the flow field, and we must develop in our later analysis an expression relating the two quantities. Nevertheless, it must be remembered that the physical mechanism of heat transfer at the wall is a conduction process.

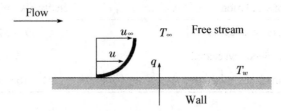

Figure 3.7 Convection heat transfer from a plate

To express the overall effect of convection, we use Newton's law of cooling

$$q = hA(T_w - T_\infty) \qquad (3-19)$$

Here the heat-transfer rate is related to the overall temperature difference between the wall and fluid and the surface area A. The quantity h is called the convection heat-transfer coefficient, and the equation (3 - 19) is the defining equation. An analytical calculation of h may be made for some systems. For complex situations it must be determined experimentally. The heat-transfer coefficient is sometimes called the film conductance because of its relation to the conduction process in the thin stationary layer of fluid at the wall surface. From the equation (3 - 19) we note that the units of h are in watts per square meter per Celsius degree when the heat flow is in watts.

In view of the foregoing discussion, one may anticipate that convection heat transfer will have a dependence on the viscosity of the fluid in addition to its dependence on the thermal properties of the fluid (thermal conductivity, specific heat and density). This is expected because viscosity influences the velocity profile and, correspondingly, the energy-transfer rate in the region near the wall.

If a heated plate were exposed to ambient room air without an external source of motion, a movement of the air would be experienced as a result of the density gradients near the plate. We call this natural, or free, convection as opposed to forced convection, which is experienced in the case of the fan blowing air over a plate. Boiling and condensation phenomena are also grouped under the general subject of convection heat transfer. The approximate ranges of convection heat-transfer coefficients are indicated in Table 3.3.

Table 3.3 Approximate values of convection heat-transfer coefficients

Mode	h	
	W / m² · ℃	Btu / h · ft² · °F
Across 2.5-cm air gap evacuated to a pressure of 10^{-6} atm and subjected to $\Delta T = 100 \sim 30$ ℃	0.087	0.015
Free convection, $\Delta T = 30$ ℃		
Vertical plate 0.3 m [1ft] high in air	4.5	0.79
Horizontal cylinder, 5-cm diameter, in air		
Horizontal cylinder, 2-cm diameter, in water	890	157
Heat transfer across 1.5-cm vertical air gap with $\Delta T = 60$ ℃	2.64	0.46
Fine wire in air, $d = 0.02$ mm, $\Delta T = 55$ ℃	490	86
Free convection		
Airflow at 2 m / s over 0.2-m square plate	12	2.1
Airflow at 35 m / s over 0.75-m square plate	75	13.2
Airflow at Mach number = 3, $p = 1/20$ atm, $T_\infty = -40$ ℃, across 0.2-m square plate	56	9.9
Air at 2 atm flow in 2.5-cm-diamter tube at 10 m / s	65	11.4
Water at 0.5 kg / s flowing in 2.5-cm-diameter tube	3 500	616
Airflow across 5-cm diameter cylinder with velocity of 50 m / s	180	32
Liquid bismuth at 4.5 kg / s and 420 ℃ in 5.0-cm-diameter tube	3 410	600
Airflow at 50 m / s across fine wire, $d = 0.04$ mm	3 850	678
Boiling water		
In a pool or container	2 500~35 000	440~6 200
Flowing in a tube	5 000~100 000	880~17 600
Condensation of water vapor, 1 atm		
Vertical surfaces	4 000~11 300	700~2 000
Outside horizontal tubes	9 500~25 000	1 700~4 400
Dropwise condensation	170 000~290 000	30 000~50 000

The energy transfer expressed by the equation (3-19) is used for evaluating the convection loss for flow over an external surface. Of equal importance is the convection gain or loss resulting from a fluid flowing inside a channel or tube as shown in Figure 3.8. In this case, the heated wall at T_w loses heat to the cooler fluid, which consequently rises in temperature as it flows from inlet conditions at T_i to exit conditions at T_e. Using the symbol i to designate enthalpy (to avoid confusion with h, the convection coefficient), the energy balance on the fluid is

$$q = \dot{m}(i_e - i_i) \tag{3-20}$$

where $\dot{m}$ is the fluid mass flow rate. For many single-phase liquids and gases operating over reasonable temperature ranges $\Delta i = c_p \Delta T$ and we have

$$q = \dot{m} c_p (T_e - T_i) \tag{3-21}$$

which may be equated to a convection relation like the equation (3-19)

$$q = \dot{m} c_p (T_e - T_i) = hA(T_{w,\text{avg}} - T_{\text{fluid,avg}}) \tag{3-22}$$

In this case, the fluid temperatures T_e, T_i and T_{fluid} are called bulk or energy average temperatures. A is the surface area of the flow channel in contact with the fluid. For now, we simply want to alert the reader to the distinction between the two types of flows.

We must be careful to distinguish between the surface area for convection that is employed in convection the equation (3-19) and the cross-sectional area that is used to calculate the flow rate from

$$\dot{m} = \rho \mu_{\text{mean}} A_c \tag{3-23}$$

where $A_c = \pi d^2/4$ for flow in a circular tube. The surface area for convection in this case would be πdL, where L is the tube length. The surface area for convection is always the area of the heated surface in contact with the fluid.

3.5 Radiation Heat Transfer

In contrast to the mechanisms of conduction and convection, where energy transfer through a material medium is involved heat may also be transferred through regions where a perfect vacuum exists. The mechanism in this case is electromagnetic radiation. We shall limit our discussion to electromagnetic radiation that is propagated as a result of a temperature difference; this is called thermal radiation.

Thermodynamic considerations show that an ideal thermal radiator, or blackbody, will emit energy at a rate proportional to the fourth power of the absolute temperature of the body and directly proportional to its surface area. Thus

$$q_{\text{emitted}} = \sigma A T^4 \tag{3-24}$$

where σ is the proportionality constant and is called the Stefan-Boltzmann constant[3] with the value of 5.669×10^{-8} W/m² · K⁴. The equation (3-24) is called the Stefan-Boltzmann law of thermal radiation, and it applies only to blackbodies. It is important to note that this equation is valid only for thermal radiation; other types of electromagnetic radiation may not be treated so simply.

The equation (3-24) governs only radiation emitted by a blackbody. The net radiant exchange between two surfaces will be proportional to the difference in absolute temperatures to the fourth power; i.e,

$$\frac{q_{\text{net exchange}}}{A} \propto \sigma (T_1^4 - T_2^4) \tag{3-25}$$

We have mentioned that a blackbody is a body that radiates energy according to the T^4 law. We call such a body black because black surfaces, such as a piece of metal covered with carbon black, approximate this type of behavior. Other types of surfaces, such as a glossy painted surface or a polished metal plate, do not radiate as much energy as the blackbody; however, the total radiation emitted by these bodies still generally follows the T^4 proportionality. To take account of the "gray" nature of such surfaces we introduce another factor into the equation (3-24), called the emissivity ϵ, which relates the radiation of the "gray" surface to that of an ideal black surface. In addition, we must take into account the fact that not all the radiation leaving one surface will reach the other surface since electromagnetic radiation travels in straight lines and some will be lost to the surroundings. We therefore introduce two new factors in the equation (3-24) to take into account both situations, so that

$$q = F_\epsilon F_G \sigma (T_1^4 - T_2^4) \tag{3-26}$$

where F_ϵ is the emissivity function, and F_G is the geometric "view factor" function. The fact that these functions usually are not independent of one another as indicated in the equation (3-26).

A simple radiation problem is encountered when we have a heat-transfer surface at temperature T_1 completely enclosed by a much larger surface maintained at T_2. The net radiant exchange in this case can be calculated with

$$q = \epsilon_1 \sigma A_1 (T_1^4 - T_2^4) \tag{3-27}$$

Radiation heat-transfer phenomena can be exceedingly complex, and the calculations are seldom as simple as implied by the equation (3-26). For now, we wish to emphasize the difference in physical mechanism between radiation heat-transfer and conduction-convection systems.

Notes

① Fourier's law 傅立叶定律　　　　② Taylor-series expansion 泰勒级数

Chapter 3 Foundation of Heat Transfer

展开

③ Stefan-Boltzmann constant 斯蒂芬-玻尔兹曼常数

◆ Vocabulary

calculus [ˈkælkjʊləs] n. 微积分学
cartesian [kɑːˈtɪzɪən] adj. 笛卡尔的
Celsius [ˈselsɪəs] n. 摄氏度 adj. 摄氏的
conduction [kənˈdʌkʃ(ə)n] n. [生理]传导
convect [kənˈvekt] vt. 使热空气对流 vi. 对流传热
convection [kənˈvekʃ(ə)n] n. [流][气象]对流；传送
cylindrical [sɪˈlɪndrɪkəl] adj. 圆柱形的；圆柱体的
derivation [derɪˈveɪʃ(ə)n] n. 引出；来历；词源；派生词
derivative [dɪˈrɪvətɪv] n. [化学]衍生物，派生物；导数 adj. 派生的；引出的
diffusivity [ˌdɪfjuːˈsɪvətɪ] n. 扩散率；[物]扩散性

emissivity [ˌemɪˈsɪvətɪ] n. [物]发射率；辐射系数
equilibrium [ˌiːkwɪˈlɪbrɪəm] n. 均衡；平静；保持平衡的能力
gradient [ˈɡreɪdɪənt] n. [数][物]梯度；坡度；倾斜度 adj. 倾斜的；步行的
momentum [məʊˈmentəm] n. 势头；动力；(物理)动量
nonideal [ˌnɒnaɪˈdɪəl] adj. 非理想的，不理想的
pail [peɪl] n. 桶，提桶
radiation [reɪdɪˈeɪʃ(ə)n] n. 辐射，放射物，辐射
thermodynamics [ˌθɜːməʊdaɪˈnæmɪks] n. 热力学
vibration [vaɪˈbreɪʃ(ə)n] n. 振动；犹豫；心灵感应

◆ Exercises

1. Translate the following Chinese phrases into English

(1) 传热 (2) 传导传热 (3) 高温区
(4) 热扩散率 (5) 绝对温度 (6) 自由流
(7) 自然对流 (8) 单相液体

2. Translate the following English phrases into Chinese

(1) hot steel bar (2) temperature gradient
(3) thermal conductivity (4) heat generation
(5) critical pressure (6) energy-transfer rate
(7) radiation heat transfer

3. Translate the following Chinese sentences into English

(1) 在绝热过程中，不容许热量进入或离开系统。
(2) 对流换热微分方程组包括质量守恒、动量守恒和能量守恒这三大定律。
(3) 大多数的热辐射体都是选择性辐射体。

（4）毛纤维热传导性很低,热量不易散失。

（5）在高温燃烧环境下,气体辐射在总的热量传递中占有不可忽略的比重。

（6）翅片越高,热量传递的路径就越长。

（7）在边界条件中考虑了铸型外表面的对流及辐射换热。

4. Translate the following English sentences into Chinese

(1) The second law of thermodynamics states that thermal transfer occurs in the direction of decreasing temperature.

(2) For the geometry condition of the convective heat transfer problems, it refers to the heat transfer geometry shape, size, position and surface roughness.

(3) It's also an excellent heat conductor, and is transparent and flexible.

(4) This kind of expansion, made without adding heat, is called an adiabatic expansion.

(5) With the rapid development of modern technologies, heat transfer phenomena are emerging in such industries as energy, aerospace, electronics, nuclear powder, *et al*.

(6) Accelerating the train speed will increase convection and cut down the influence of radiation so as to lower the integrated temperature of walls.

5. Translate the following Chinese essay into English

物体各部分之间不发生相对位移时,依靠分子、原子及自由电子等微观粒子的热运动而产生的热量传递称为导热。学习导热的目标主要有两个:①传导了多少热量?②固体内部的温度怎么分布? 前者由傅立叶导热定律决定,后者主要由导热微分方程进行求解。

6. Translate the following English essay into Chinese

This chapter introduces the Newton's law of cooling, convective heat differential equations and factors of convective heat. Secondly, it introduces the boundary layer theory, and according to the conservation of mass, conservation of energy and momentum conservation to get a set of differential equations of convective heat and simplified set of differential equations boundary. Then, describe the similarity principle and dimensional analysis of theory and application of theory to obtain guidelines correlation. Finally, a typical tube flow tube flow and natural convection, the application of the test criteria correlation be solved in order to obtain the surface convective heat transfer coefficient.

扫一扫,查看更多资料

Chapter 4　Diffusion Theory

4.1　Introduction

Everybody is familiar with the experience that a drop of ink slowly dissolves in water or smoke spreads out in air. The reason for these phenomena is the motion of the molecules in a liquid or a gas. Although less obvious, atoms in a solid are also capable of leaving their lattice sites by thermal activation and to move through the crystal. This is referred to as solid state diffusion.

To begin with, we want to emphasize that diffusion is a process which is not due to the action of a specific force but rather that diffusion is a result of the random movement of atoms, i.e., and a statistical problem. For a most general treatment of the problem we will first disregard the atomic nature of the diffusing particles and define their amount by the concentration c, i.e., the number of particles per unit volume $[cm^{-3}]$.

Many reactions and processes that are important in the treatment of materials rely on the transfer of mass either within a specific solid (ordinarily on a microscopic level) or from a liquid, a gas or another solid phase. This is necessarily accomplished by diffusion, the phenomenon of material transport by atomic motion. This chapter discusses the atomic mechanisms by which diffusion occurs, the mathematics of diffusion, and the influence of temperature and diffusing species on the rate of diffusion.

The phenomenon of diffusion may be demonstrated with the use of a diffusion couple, which is formed by joining bars of two different metals together so that there is intimate contact between the two faces; this is illustrated for copper and nickel in Figure 4.1, which includes schematic representations of atom positions and composition across the interface. This couple is heated for an extended period at an elevated temperature (but below the melting temperature of both metals), and cooled to room temperature. Chemical analysis will reveal a condition similar to that represented in Figure 4.2, namely, pure copper and nickel at the two extremities of the couple, separated by an alloyed region. Concentrations of both metals vary with position as shown in Figure 4.2c. This result indicates that copper atoms have migrated or diffused into the nickel, and that nickel has diffused into copper. This process, whereby atoms of one metal diffuse into another, is termed inter diffusion, or impurity diffusion.

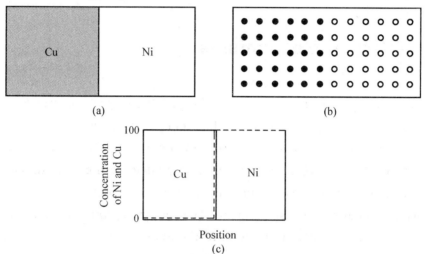

Figure 4.1 A copper-nickel diffusion couple before a high-temperature heat treatment (a), schematic representations of Cu (solid circles) and Ni (vacuous circles) atom locations within the diffusion couple (b) and concentrations of copper and nickel as a function of position across the couple (c)

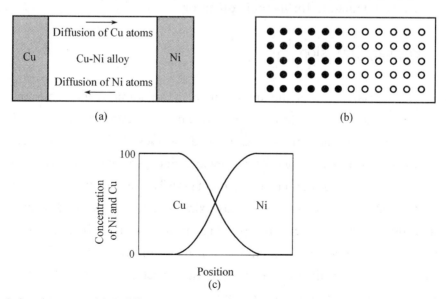

Figure 4.2 A copper-nickel diffusion couple after a high-temperature heat treatment, showing the alloyed diffusion zone (a), schematic representations of Cu (solid circles) and Ni (vacuous circles) atom locations within the couple (b) and concentrations of copper and nickel as a function of position across the couple (c)

Chapter 4 Diffusion Theory

Inter diffusion may be discerned from a macroscopic perspective by changes in concentration which occur over time, as in the example for the Cu-Ni diffusion couple. There is a net drift or transport of atoms from high to low concentration regions. Diffusion also occurs for pure metals, but all atoms exchanging positions are of the same type; this is termed sclf-diffusion. Of course, self-diffusion is not normally subject to observation by noting compositional changes.

4.2 Diffusion Mechanisms

From an atomic perspective, diffusion is just the stepwise migration of atoms from lattice site to lattice site. In fact, the atoms in solid materials are in constant motion, rapidly changing positions. For an atom to make such a move, two conditions must be met: there must be an empty adjacent site, and the atom must have sufficient energy to break bonds with its neighbor atoms and then cause some lattice distortion during the displacement. This energy is vibrational in nature. At a specific temperature some small fraction of the total number of atoms is capable of diffusive motion, by virtue of the magnitudes of their vibrational energies. This fraction increases with rising temperature.

Several different models for this atomic motion have been proposed; of these possibilities, two dominate for metallic diffusion.

4.2.1 Vacancy Diffusion

One mechanism involves the interchange of an atom from a normal lattice position to an adjacent vacant lattice site or vacancy, as represented schematically in Figure 4.3a. This mechanism is aptly termed vacancy diffusion. Of course, this process necessitates the presence of vacancies, and the extent to which vacancy diffusion can occur is a function of the number of these defects that are present; significant concentrations of vacancies may exist in metals at elevated temperatures. Since diffusing atoms and vacancies exchange positions, the diffusion of atoms in one direction corresponds to the motion of vacancies in the opposite direction. Both self-diffusion and inter-diffusion occur by this mechanism; for the latter, the impurity atoms must substitute for host atoms.

Chapter 4　Diffusion Theory

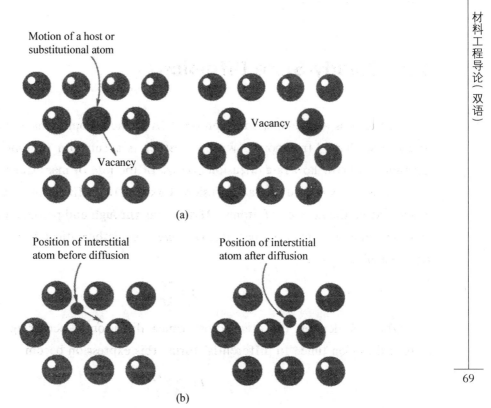

Figure 4.3　Schematic representations of vacancy diffusion (a) and interstitial diffusion (b)

4.2.2　Interstitial Diffusion

The second type of diffusion involves atoms that migrate from an interstitial position to a neighboring one that is empty. This mechanism is found for interdiffusion of impurities such as hydrogen, carbon, nitrogen and oxygen, which have atoms that are small enough to fit into the interstitial positions. Host or substitution impurity atoms rarely form interstitials and do not normally diffuse via this mechanism. This phenomenon is appropriately termed interstitial diffusionin Figure 4.3b.

In most metal alloys, interstitial diffusion occurs much more rapidly than diffusion by the vacancy mode, since the interstitial atoms are smaller, and thus more mobile. Furthermore, there are more empty interstitial positions than vacancies; hence, the probability of interstitial atomic movement is greater than for vacancy diffusion.

Chapter 4 Diffusion Theory

4.3 Steady-State Diffusion

Diffusion is a time-dependent process[①] in a macroscopic sense; the quantity of an element that is transported within another is a function of time. Often it is necessary to know how fast diffusion occurs, or the rate of mass transfer. This rate is frequently expressed as a diffusion flux (J), defined as the mass (or, equivalently, the number of atoms) M diffusing through and perpendicular to a unit cross-sectional area of solid per unit of time. In mathematical form, this may be represented as

$$J = \frac{M}{At} \qquad (4-1)$$

Where A denotes the area across which diffusion is occurring and t is the elapsed diffusion time. In differential form, this expression becomes

$$J = \frac{1}{A}\frac{dM}{dt} \qquad (4-2)$$

The units for J are kilograms or atoms per meter squared per second (kg · m^{-2} · s^{-1} or atoms · m^{-2} · s^{-1}).

If the diffusion flux does not change with time, a steady-state condition exists. One common example of steady-state diffusion is the diffusion of atoms of a gas through a plate of metal for which the concentrations (or pressures) of the diffusing species on both surfaces of the plate are held constant. This is represented schematically in Figure 4.4a.

When concentration C is plotted versus position (or distance) within the solid x, the resulting curve is termed the concentration profile; the slope at a particular point on this curve is the concentration gradient:

$$\text{concentration gradient} = \frac{dC}{dx} \qquad (4-3)$$

In the present treatment, the concentration profile is assumed to be linear, as depicted in Figure 4.4b, and

$$\text{concentration gradient} = \frac{\Delta C}{\Delta x} = \frac{C_A - C_B}{x_A - x_B} \qquad (4-4)$$

For diffusion problems, it is sometimes convenient to express concentration in terms of mass of diffusing species per unit volume of solid (kg · m^{-3} or g · cm^{-3}).

Chapter 4 Diffusion Theory

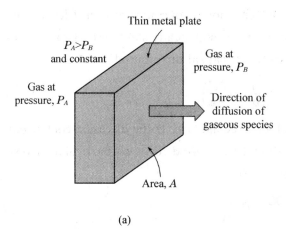
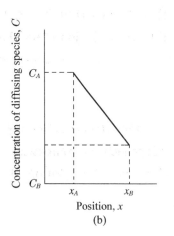

Figure 4.4 (a) Steady-state diffusion across a thin plate and (b) a linear concentration profile for the diffusion situation in (a)

The mathematics of steady-state diffusion in a single (x) direction is relatively simple, in that the flux is proportional to the concentration gradient through the expression

$$J = -D \frac{dC}{dx} \qquad (4-5)$$

The constant of proportionality D is called the diffusion coefficient, which is expressed in square meters per second. The negative sign in this expression indicates that the direction of diffusion is down the concentration gradient, from a high to a low concentration. The equation (4-5) is sometimes called Fick's first law[2].

Sometimes the term driving force is used in the context of what compels a reaction to occur. For diffusion reactions, several such forces are possible; but when diffusion is according to the equation (4-5), the concentration gradient is the driving force.

One practical example of steady-state diffusion is found in the purification of hydrogen gas. One side of a thin sheet of palladium metal is exposed to the impure gas composed of hydrogen and other gaseous species such as nitrogen, oxygen and water vapor. The hydrogen selectively diffuses through the sheet to the opposite side, which is maintained at a constant and lower hydrogen pressure.

4.4 Nonsteady-State Diffusion

Most practical diffusion situations are nonsteady-state ones. That is, the diffusion flux and the concentration gradient at some particular point in a solid vary with time, with a net accumulation or depletion of the diffusing species resulting.

Chapter 4 Diffusion Theory

This is illustrated in Figure 4.5, which shows concentration profiles at three different diffusion times. Under conditions of non-steady state, use of the equation (4-5) is no longer convenient; instead, the partial differential equation

$$\frac{\partial C}{\partial t} = \frac{\partial}{\partial x}\left(D \frac{\partial C}{\partial x}\right) \tag{4-6}$$

The above equation is known as Fick's second law. If the diffusion coefficient is independent of composition (which should be verified for each particular diffusion situation), the equation (4-6) simplifies to

$$\frac{\partial C}{\partial t} = D \frac{\partial^2 C}{\partial x^2} \tag{4-7}$$

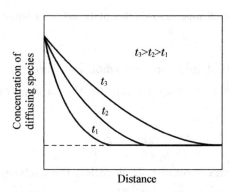

Figure 4.5 Concentration profiles for nonsteady-stae diffusion taken at three different times t_1, t_2 and t_3

Solutions to this expression (concentration in terms of both position and time) are possible when physically meaningful boundary conditions are specified.

One practically important solution is for a semi-infinite solid in which the surface concentration is held constant. Frequently, the source of the diffusing species is a gas phase, the partial pressure, of which is maintained at a constant value. Furthermore, the following assumptions are made:

(1) Before diffusion, any of the diffusing solute atoms in the solid are uniformly distributed with concentration of C_0.

(2) The value of x at the surface is zero and increases with distance into the solid.

(3) The time is taken to be zero the instant before the diffusion process begins.

These boundary conditions are simply stated as

For $t = 0$, $C = C_0$ at $0 \leqslant x \leqslant \infty$.

For $t > 0$, $C = C_s$ (the constant surface concentration) at $x = 0$.

$C = C_0$ at $x = \infty$.

Application of these boundary conditions to the equation (4-7) yields the solution

$$\frac{C_x - C_0}{C_s - C_0} = 1 - erf\left(\frac{x}{2\sqrt{Dt}}\right) \quad (4-8)$$

where C_x represents the concentration at depth x after time t. The expression $erf(x/2\sqrt{Dt})$ is the Gaussian error function[③]. This Gaussian error function is defined by

$$erf(x) = \frac{2}{\pi} \int_0^x e^{-y^2} dy \quad (4-9)$$

where $x/2\sqrt{Dt}$ has been replaced by the variable z.

The concentration parameters that appear in the equation (4-8) are noted in Figure 4.6, a concentration profile taken at a specific time. The equation (4-8) thus demonstrates the relationship among concentration, position and time, namely, that C_x, being a function of the dimensionless parameter $x/\sqrt{Dt}$, may be determined at any time and position if the parameters C_0, C_s and D are known.

Suppose that it is desired to achieve some specific concentration of solute, C_1, in an alloy; the left-hand side of the equation (4-8) now becomes

$$\frac{C_1 - C_0}{C_s - C_0} = \text{constant} \quad (4-10)$$

This being the case, the right-hand side of this same expression is also a constant, and subsequently

$$\frac{x}{2\sqrt{Dt}} = \text{constant} \quad (4-11)$$

Or

$$\frac{x^2}{Dt} = \text{constant} \quad (4-12)$$

Some diffusion computations are thus facilitated on the basis of this relationship.

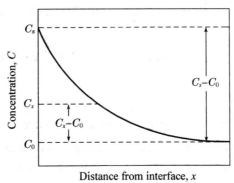

Figure 4.6 Concentration profile for nonsteady-state diffusion; concentration parameters relate to the equation (4-8)

Chapter 4 Diffusion Theory

4.5 Factors That Influence Diffusion

4.5.1 Diffusing Species

The magnitude of the diffusion coefficient D is indicative of the rate at which atoms diffuse coefficients, both self- and inter-diffusion. The diffusing species as well as the host material influence the diffusion coefficient. For example, there is a significant difference in magnitude between self- and carbon inter-diffusion in an iron at 500 ℃, the D value being greater for the carbon inter-diffusion (3.0×10^{-21} vs. 2.4×10^{-12} m² · s⁻¹). This comparison also provides a contrast between rates of diffusion via vacancy and interstitial modes as discussed above. Self-diffusion occurs by a vacancy mechanism, whereas carbon diffusion in iron is interstitial.

4.5.2 Temperature

Temperature has a most profound influence on the coefficients and diffusion rates. For example, for the self-diffusion of Fe in α-Fe, the diffusion coefficient increases approximately six orders of magnitude (from 3.0×10^{-21} to 1.8×10^{-15} m² · s⁻¹) in rising temperature from 500 to 900 ℃. The temperature dependence of diffusion coefficients is related to temperature according to

$$D = D_0 \exp\left(-\frac{Q_d}{RT}\right) \quad (4-13)$$

Where

D_0 = a temperature-independent pre-exponential (m² · s⁻¹).

Q_d = the activation energy for diffusion (J · mol⁻¹, cal · mol⁻¹, or eV · atom⁻¹).

R = the gas constant, 8.31 J · mol⁻¹ · K⁻¹, 1.987 cal · mol⁻¹ · K⁻¹, or 8.62×10^{-5} eV · atom⁻¹.

T = absolute temperature (K).

The activation energy may be thought of as that energy required producing the diffusive motion of one mole of atoms. Large activation energy results in a relatively small diffusion coefficient. We take D_0 and Q_d values for several diffusion systems for example. Taking natural logarithms of the equation (4-13) yields

$$\ln D = \ln D_0 - \frac{Q_d}{R}\left(\frac{1}{T}\right) \tag{4-14}$$

Or in terms of logarithms to the base 10.

Since D_0, Q_d and R are all constants, the equation (4-14) takes on the form of an equation straight line:

$$y = b + mx \tag{4-15}$$

where y and x are analogous, respectively, to the variables $\ln D$ and $1/T$. Thus, if $\ln D$ is plotted versus the reciprocal of the absolute temperature, a straight line should result, having slope and intercept of $-Q_d/2.3R$ and $\log D_0$, respectively. This is, in fact, the manner in which the values of Q_d and D_0 are determined experimentally.

Notes

① time-dependent process 含时过程
② Fick's first law 菲克第一定律
③ Gaussian error function 高斯误差函数

Vocabulary

analogous [əˈnæləgəs] *adj.* 类似的
diffusion [dɪˈfjuːʒ(ə)n] *n.* 扩散, 传播; [光]漫射
dimensionless [dɪˈmenʃənlɪs] *n.* 无穷小量 *adj.* [物]无量纲的; [数]无因次的; 无尺寸的
impurity [ɪmˈpjʊərɪtɪ] *n.* 杂质; 不纯; 不洁

interstitial [ˌɪntəˈstɪʃ(ə)l] *adj.* 间质的; 空隙的; 填隙的 *n.* 填隙原子
logarithm [ˈlɒgərɪð(ə)m] *n.* [数]对数
palladium [pəˈleɪdɪəm] *n.* [化学]钯
profile [ˈprəʊfaɪl] *n.* 侧面; 轮廓; 外形; 剖面; 简况 *vt.* 描……的轮廓; 扼要描述 *vi.* 给出轮廓

Exercises

1. Translate the following Chinese phrases into English

(1) 扩散理论　　　(2) 熔化温度　　　(3) 置换原子
(4) 微分形式　　　(5) 薄金属板　　　(6) 菲克第二定律
(7) 气体常数　　　(8) 活化能　　　　(9) 直线

2. Translate the following English phrases into Chinese

(1) solid phase　　　　　　(2) vacancy diffusion
(3) interstitial position　　　(4) concentration profile
(5) concentration gradient　　(6) semi-infinite solid
(7) dimensionless parameter　(8) natural logarithm

Chapter 4 Diffusion Theory

3. Translate the following Chinese sentences into English

（1）溶液中的气体会由浓度高的部分扩散到浓度低的部分。

（2）相对尺寸较小的溶质原子占据溶剂或晶格原子之间间隙位置所形成的固溶体。

（3）上述的许多问题可用稳态扩散理论加以推理。

（4）非牛顿流体是普遍存在的，而目前扩散理论的研究仅限于牛顿流体。

（5）平衡吸附量随溶液浓度的增大而提高。

（6）纯铁渗氮层则由化合物层和扩散层组成。

4. Translate the following English sentences into Chinese

（1）The diffusion flux is proportional to the concentration gradient.

（2）The order, the velocity constants and the apparent activation energy were obtained.

（3）The density of an ideal gas at constant pressure is in direct proportion to the temperature.

（4）The oxygen reacts vigorously with the impurity in the iron.

（5）However, at high temperature, when the matrix is in the state of low saturation, precipitations in grain and on grain boundary are controlled by thermodynamics.

5. Translate the following Chinese essay into English

通过控制相变或固体材料的反应，固相反应为生产纳米粒子提供了可能性。这种方法的优点是生产工艺简单。固相反应技术是用于高产率地生产单分散的纳米氧化物纳米粒子的一种便捷的、廉价的及高效的制备方法。

这是制备许多陶瓷粉末的最传统的方式，如：$BaTiO_3$。它涉及将 $BaCO_3$ 和 TiO_2 颗粒进行粉磨以进行粉碎和混合。然后，将混合物在 900～1 200 ℃下煅烧使其生成 $BaTiO_3$。反应机理如下。

（1）Ba 通过扩散进入 TiO_2 中发生界面反应，在 TiO_2 周围形成 $BaTiO_3$ 壳（层）；生成的壳限制了 Ba 向 TiO_2 中进一步扩散。

（2）形成原钛酸盐相。

（3）最终形成 $BaTiO_3$。

高的焙烧温度以促进 Ba 和 Ti 的相互扩散，导致固态反应形成 $BaTiO_3$。

6. Translate the following English essay into Chinese

Nitriding is also one of the case hardening methods. This process consists of keeping the steel in hot ammonia gas for some hours. Nitrogen, formed in this condition from ammonia, penetrates into the surface of the metal, thus forming a very hard case.

扫一扫，查看更多资料

Part II
Design and Simulation of Materials

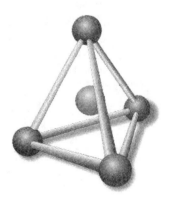

Part II
Design and Simulation of Materials

Chapter 5 Computer-Aided Design and Manufacturing of Materials

5.1 Introduction

5.1.1 Material Forming and Manufacturing

As you begin to read this introduction, take a few moments and inspect the different objects around you: your watch, chair, stapler, pencil, calculator, telephone and light fixtures. You will soon realize that all these objects have different shapes at one time. You could not find them in nature as they appear in your room. They have been transformed from various raw materials and assembled into the shapes that you now see.

Some objects are made of one part, such as nails, bolts, wire or plastic coat hangers, metal brackets and forks. However, most objects aircraft engines, ballpoint pens, toasters, bicycles, computers and thousands more are made of a combination of several parts made from a variety of materials. A typical automobile for example, consists of about 15 000 parts, a C – 5A transport plane is made of more than four million parts, and a Boeing 747 – 400 is made of six million parts. All are made by various processes that we call manufacturing. Manufacturing, in its broadest sense, is the process of converting raw materials into products. It encompasses the design and manufacturing of goods, using various production, methods and techniques.

Manufacturing is the backbone of any industrialized nation. Its importance is emphasized by the fact that as an economic activity, it comprises approximately 22 to 30 percent of the value of all goods and services produced in industrialized nations.

The level of manufacturing activity is directly related to the economic health of

Chapter 5 Computer-Aided Design and Manufacturing of Materials

a country. Generally, the higher the level of manufacturing activity in a country, the higher is the standard of living of its people.

Manufacturing also involves activities in which the manufactured product is itself used to make other products. Examples are large presses to form sheet metal for car bodies, metalworking machinery used to make parts for other products, and sewing machines for making clothing. An equally important aspect of manufacturing activities is servicing and maintaining this machinery during its useful life.

The word manufacturing is derived from the Latin manu factus, meaning made by hand. The word manufacture first appeared in 1567, and the word manufacturing appeared in 1683. In the modern sense, manufacturing involves making products from raw materials by various processes, machinery and operations, following a well organized plan for each activity required. The word product means something that is produced, and the words product and production first appeared sometime during the fifteenth century. The word production is often used interchangeably with the word manufacturing. Whereas manufacturing engineering is the term used widely in the United States to describe this area of industrial activity, the equivalent term in Europe and Japan is production engineering.

Because a manufactured item has undergone a number of changes in which a piece of raw material has become a useful product, it has a value defined as monetary worth or marketable price. For example, as the raw material for ceramics, clay has a certain value as mined. When the clay is used to make a ceramic dinner plate, cutting tool or electrical insulator, value is added to the clay. Similarly, a wire coat hanger or a nail has a value over and above the cost of a piece of wire. Thus manufacturing has the important function of adding value.

Manufacturing may produce discrete products, meaning individual parts or part pieces, or continuous products. Nails, gears, steel balls, beverage cans and engine blocks are examples of discrete parts, even though they are mass produced at high rates. On the other hand, a spool of wire, metal or plastic sheet, tubes, hose and pipe are continuous products, which may be cut into individual pieces and thus become discrete parts.

Manufacturing is generally a complex activity, involving people who have a broad range of disciplines and skills and a wide variety of machinery, equipment and tooling with various levels of automation, including computers, robots and material handling equipment. Manufacturing activities must be responsive to several demands and trends:

Chapter 5 Computer-Aided Design and Manufacturing of Materials

(1) A product must fully meet design requirements and specifications.

(2) A product must be manufactured by the most economical methods in order to minimize costs.

(3) Quality must be built into the product at each stage, from design to assembly, rather than relying on quality testing after the product is made.

(4) In a highly competitive environment, production methods must be sufficiently flexible so as to respond to changing market demands, types of products, production rates, production quantities, and on time delivery to the customer.

(5) New developments in materials, production methods and computer integration of both technological and managerial activities in a manufacturing organization must constantly be evaluated with a view to their timely and economic implementation.

(6) Manufacturing activities must be viewed as a large system, each part of which is interrelated to others. Such systems can be modeled in order to study the effect of factors such as changes in market demands, product design, material and various other costs, and production methods on product quality and cost.

(7) The manufacturing organization must constantly strive for higher productivity, defined as the optimum use of all its resources: materials, machines, energy, capital, labor and technology. Output per employee per hour in all phases must be maximized.

5.1.2 Design Process and Concurrent Engineering

The design process for a product first requires a clear understanding of the functions and the performance expected of that product (Figure 5.1). The product may be new, or it may be a revised version of an existing product. We all have observed, for example, how the design and style of radios, toasters, watches, automobiles and washing machines have changed. The market for a product and its anticipated uses must be defined clearly, with the assistance of sales personnel, market analysts and others in the organization. Product design is a critical activity because it has been estimated that 70 to 80 percent of the cost of product development and manufacture is determined at the initial design stages.

Chapter 5 Computer-Aided Design and Manufacturing of Materials

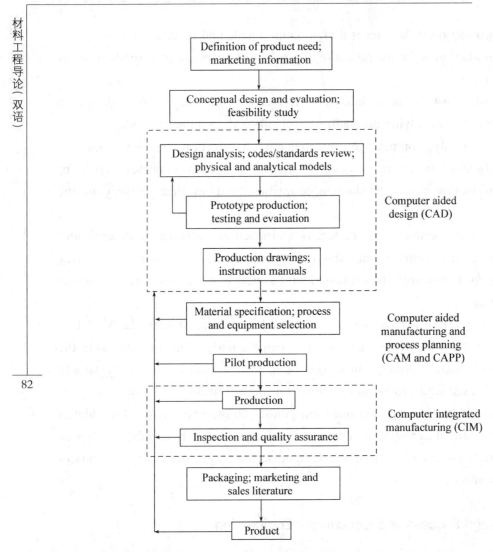

Figure 5.1 Chart showing various steps involved in designing and manufacturing a product. Depending on the complexity of the product and the type of materials used, the time span between the original concept and marketing a product may range from a few months to many years. Concurrent engineering combines these stages to reduce the time span and improve efficiency and productivity

Traditionally, design and manufacturing activities have taken place sequentially rather than concurrently or simultaneously. Manufacturing engineers were given the detailed drawings and specifications of the product and were asked to make them. They often encountered difficulties because the design or product engineers did not anticipate the production problems which could occur. This situation has been improved greatly by the use of concurrent engineering (CE) or simultaneous engineering.

The design process begins with the development of an original product concept. An innovative approach to design is highly desirable and even essential at this stage

Chapter 5　Computer-Aided Design and Manufacturing of Materials

for the product to be successful in the marketplace. Innovative approaches can also lead to major savings in material and production costs. The design engineer or product designer must be knowledgeable of the interrelationships among materials, design and manufacturing, as well as the overall economics of the operation. Concurrent engineering is a systematic approach integrating the design and manufacture of products with the view of optimizing all elements involved in the life cycle of the product. Life cycle means that all aspects of a product, such as design, development, production, distribution, use, and its ultimate disposal and recycling, are considered simultaneously. The basic goals of CE are to minimize product design and engineering changes and the time and costs involved in taking the product from design concept to production and introduction of the product into the marketplace. An extension of CE called direct engineering (DE) has recently been proposed. DE utilizes a database representing the engineering logic used in, the design of each part of a product. If a design modification is made on a part, DE will determine the manufacturing consequences of that change.

Although the concept of concurrent engineering appears to be logical and efficient, its implementation can take considerable time and effort. When those using it either do not work as a team or fail to appreciate its real benefits, It is apparent that for CE to succeed it must have the full support of the upper management, have a multifunctional and interactive teamwork, including support groups, and utilize all available technologies.

There are many examples of the benefits of concurrent engineering. An automotive company, for example, has reduced the number of components in an engine by 30 percent, decreasing its weight by 25 percent and cutting manufacturing time by 50 percent. The CE concept can be implemented not only in large organizations but also in smaller companies. This is particularly noteworthy in view of the fact that 98 percent of U. S. manufacturing establishments have fewer than 500 employees.

For both large and small companies, product design often involves preparing analytical and physical models of the product, as an aid to analyzing factors such as forces, stresses, deflections and optimal part shape. The necessity for such models depends on product complexity. Today, constructing and studying analytical models is simplified through the use of computer aided design, engineering and manufacturing techniques. On the basis of these models, the product designer selects and specifies the final shape and dimensions of the product, its surface finish and dimensional accuracy, and the materials to be used. The selection of materials is often made with the advice and cooperation of materials engineers, unless the design engineer is also experienced and qualified in this area.

An important design consideration is how a particular component is to be assembled into the final product. Take apart a ballpoint pen or a toaster, or lift the hood of your car and observe how hundreds of components are put together in a limited space.

The next step in the production process is to make and test a prototype, that is, an original working model of the product. An important new development is rapid prototyping, which relies on CAD/CAM and various manufacturing techniques (using polymers or metal powders) to produce prototypes in the form of a solid physical model of a part rapidly and at low cost. For example, prototyping new automotive components (by traditional methods of shaping, forming, machining, *etc.*) cost hundreds of millions of dollars a year; some components may take a year to make. Rapid prototyping can cut these costs as well as development times significantly. These techniques are being advanced further so that they cart be used for low volume economical production of parts.

Tests on prototypes must be designed to simulate as closely as possible the conditions under which the product is to be used. These include environmental conditions such as temperature and humidity, as well as the effects of vibration and repeated use and misuse of the product. Computer aided engineering techniques are now capable of comprehensively and rapidly performing such simulations. During this stage, modifications in the original design, materials selected or production methods may be necessary. After this phase has been completed, appropriate process plans, manufacturing methods, equipment and tooling are selected with the cooperation of manufacturing engineers, process planners and all others involved in production.

Overdesign: Surveys have indicated that many products in the past have been overdesigned. That is, they were either too bulky, were made of materials too high in quality, or were made with unwarranted precision and quality for the intended uses.

Overdesign may result from uncertainties in design calculations or the concern of the designer and manufacturer over product safety in order to avoid user injuries or deaths and resulting product liability lawsuits. Many designs are based on past experience and intuition, rather than on thorough analysis and experimentation. Overdesign can add significantly to the product's cost. We must point out, however, that this entire subject is somewhat controversial. From the consumer's standpoint, an automobile, washing machine or lawnmower that has been operating satisfactorily for many years without needing repairs or part replacement is a good

Chapter 5 Computer-Aided Design and Manufacturing of Materials

product.

Manufacturers are sensitive to the public image of their products in an expanding global marketplace. In fact, some products that require infrequent repair, such as washers, driers and automobiles, have been advertised as such in the public media. However, many manufacturers believe that if a product functions well for an extended period of time, it may have been overdesigned. In such cases, the company may consider downgrading the materials and/or the processes used. Some industries have even been accused of following a strategy of planned obsolescence in order to generate more sales over a period of time.

5.1.3 Design for Manufacture and Assembly

As we have seen, design and manufacturing must be intimately interrelated. Design and manufacturing should never be viewed as separate disciplines or activities. Each part or component of a product must be designed so that it not only meets design requirements and specifications, but also can be manufactured economically and with relative ease. This approach improves productivity and allows a manufacturer to remain competitive.

This broad view has now become recognized as the area of design for manufacture (DFM). It is a comprehensive approach to production of goods and integrates the design process with materials, manufacturing methods, process planning, assembly, testing and quality assurance. Effectively implementing design for manufacture requires that designers have a fundamental understanding of the characteristics, capabilities, and limitations of materials, manufacturing processes, and related operations, machinery, and equipment. This knowledge includes characteristics such as variability in machine performance, surface finish and dimensional accuracy of the workpiece, processing time, and the effect of processing method on part quality.

Designers and product engineers must be able to assess the impact of design modifications on manufacturing process selection, assembly, inspection, tools and dies, and product cost. Establishing quantitative relationships is essential in order to optimize the design with ease of manufacturing and assembly at minimum product cost (also called producibility). Computer aided design, engineering, manufacturing and process planning techniques, using powerful computer programs, have become indispensable to those conducting such analysis. New developments include expert systems, which have optimization capabilities, thus expediting the traditional iterative process in design optimization.

Chapter 5 Computer-Aided Design and Manufacturing of Materials

After individual parts have been manufactured, they are assembled into a product. Assembly is an important phase of the overall manufacturing operation and requires consideration of the ease, speed and cost of putting parts together. Also, many products must be designed so that disassembly is possible, enabling the products to be taken apart for maintenance, servicing or recycling of their components. Because assembly operations can contribute significantly to product cost, design for assembly (DFA) as well as design for disassembly is now recognized as important aspects of manufacturing. Typically, a product that is easy to assemble is also easy to disassemble.

Methodologies and computer software have been developed for DFA utilizing 3D conceptual designs and solid models. In this way, subassembly and assembly times and costs are minimized while maintaining product integrity and performance; the system also improves the product's ease of disassembly. The trend now is to combine design for manufacture and design for assembly into the more comprehensive design for manufacture and assembly (DFMA) which recognizes the inherent interrelationships between design and manufacturing.

There are several methods of assembly, each with its own characteristics and requiring different operations. The use of a bolt and nut, for example, requires preparation of holes that must match in location and size. Hole generation[①] requires operations such as drilling or punching, which take additional time, require separate operations and produce scrap. On the other hand, products assembled with bolts and nuts can be taken apart and reassembled with relative ease.

Parts can also be assembled with adhesives. This method, which is being used extensively in aircraft and automobile production, does not require holes. However, surfaces to be assembled must match properly and be clean because joint strength is affected by the presence of contaminants such as dirt, dust, oil and moisture. Unlike mechanical fastening, adhesively joined components as well as those that are welded, which are not usually designed to be taken apart and reassembled, hence are not suitable for the important purposes of recycling individual parts in the product.

Parts may be assembled by hand or by automatic equipment and robots. The choice depends on factors such as the complexity of the product, the number of parts to be assembled, the protection required to prevent damage or scratching of finished surfaces of the parts, and the relative costs of labor and machinery required for automated assembly.

Chapter 5 Computer-Aided Design and Manufacturing of Materials

5.1.4 Selecting Materials

An ever increasing variety of materials is available, each having its own characteristics, applications, advantages and limitations. The following are the general types of materials used in manufacturing today either individually or in combination:

(1) Ferrous metals (carbon, alloy, stainless, tool and die steels).

(2) Nonferrous metals and alloys (aluminum, magnesium, copper, nickel, titanium, superalloys, refractory metals, beryllium, zirconium, low melting alloys and precious metals).

(3) Plastics (thermoplastics, thermosets and elastomers).

(4) Ceramics (glass ceramics, glasses, graphite and diamond).

(5) Composite materials (reinforced plastics, metal matrix and ceramic matrix composites, and honeycomb structures). These are also known as engineered materials.

As new materials are developed, there are important trends in the selection of materials. In the example of the new bicycle, the spoke wheels are made of composite materials, the tubes are made of carbon fiber reinforced plastic, and the rear dropouts, headset and seat mast are made of titanium.

Because of the interest of the producers of different types of metallic and nonmetallic materials, there are constantly shifting trends in the usage of these materials. These trends are driven principally by economic considerations. Steel producers, for example, are countering the increased use of plastics in cars and aluminum in beverage cans. Likewise, aluminum producers are countering the use of various materials in cars. Several other such competitive applications can be cited, all of which ultimately benefit the customer.

5.1.5 Selecting Manufacturing Processes

Many processes are used to produce parts and shapes. As you can see, there is usually more than one method of manufacturing a part from a given material. The broad categories of processing methods for materials are:

(1) Casting (expendable mold and permanent mold).

(2) Forming and shaping (rolling, forging, extrusion, drawing, sheet forming, powder metallurgy and molding).

(3) Machining (turning, boring, drilling, milling, planning, shaping,

Chapter 5　Computer-Aided Design and Manufacturing of Materials

broaching, grinding, ultrasonic machining; chemical, electrical and electrochemical machining; and high energy beam machining).

(4) Joining (welding, brazing, soldering, diffusion bonding, adhesive bonding and mechanical joining).

(5) Finishing operations (honing, lapping, polishing, burnishing, deburring, surface treating, coating and plating).

Selection of a particular manufacturing process depends not only on the shape to be produced but also on a large number of other factors. The type of material and its properties are basic considerations. Brittle and hard materials, for example, cannot be shaped easily, whereas they can be cast or machined by several methods. The manufacturing process usually alters the properties of materials. Metals that are formed at room temperature, for example, become stronger, harder and less ductile than they were before processing.

Figure 5.2 show two steel mounting brackets, one made by casting, another made by stamping of sheet metal. Note that there are some differences in the designs, although the parts are basically alike. Each of these two manufacturing processes has its own advantages and limitations, as well as production rates and cost.

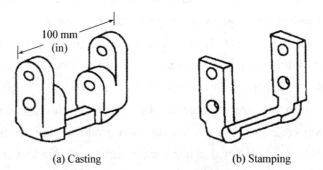

(a) Casting　　　　　　　　(b) Stamping

Figure 5.2　Two steps manufacturing process of steel mounting brackets

Manufacturing engineers are constantly being challenged to find new solutions to manufacturing problems and cost reduction. For a long time, for example, sheet metal parts were cut and made by traditional tools, punches and dies. Although they are still widely used, these operations are now being replaced by laser cutting techniques. With advances in computer controls, we can automatically control the path of the laser, thus producing a wide variety of shapes accurately, repeatedly and economically.

Ultraprecision manufacturing techniques and machinery are now being developed and are coming into more common use. For machining mirror like surfaces, for example, the cutting tool is a very sharp diamond tip and the equipment has very high stiffness and must be operated in a room where the temperature is controlled within less than 1 ℃.

5.1.6 Computer Integrated Manufacturing

The major goals of automation in manufacturing facilities are to integrate various operations to improve productivity, increase product quality and uniformity, minimize cycle times and effort, and reduce labor costs. Beginning in the 1940s, automation has accelerated because of rapid advances in control systems for machines and in computer technology.

Few developments in the history of manufacturing have had a more significant impact than computers. Beginning with computer graphics, the use of computes has been extended to computer aided design and manufacturing ultimately, to computer integrated manufacturing systems. The use of computers covers a broad range of applications, including control and optimization of manufacturing processes, material handling, assembly, automated inspection and testing of products.

5.1.7 Machine Control Systems

Numerical control (NC) of machines is a method of controlling the movements of machine components by direct insertion of coded instructions in the form of numerical data. Numerical control was first implemented in the early 1950s and was a major advance in automation of machines. With advances in computers, numerical control has expanded into computer numerical control (CNC).

In adaptive control (AC), process parameters are adjusted automatically to optimize production rate and product quality and minimize cost. Parameters such as forces, temperatures, surface finish and dimensions of the part are monitored constantly. If they move outside the acceptable range, the system adjusts the process variables until the parameters again fall within the acceptable range.

Major advances have been made in automated handling of materials in various stages of completion such as from storage to machines, from machine to machine and at the points of inspection, inventory and shipment. Introduced in the early 1960s, industrial robots are replacing humans in operations that are repetitive, boring and dangerous, thus reducing the possibility of human error, decreasing variability in product quality, and improving productivity. Robots with sensory perception capabilities are being developed, with movements that are beginning to simulate those of humans. Automated and robotic assembly systems are replacing costly assembly by operators. Products are being designed or redesigned, so that they can be assembled more easily by machine.

5.1.8 Computer Technology

Computers allow us to integrate virtually all phases of manufacturing operations, which consist of various technical as well as managerial activities. With sophisticated software and hardware, manufacturers are now able to minimize manufacturing costs, improve product quality, reduce product development time, and maintain a competitive edge in the domestic and international marketplace.

Computer integrated manufacturing (CIM) is particularly effective, with benefits including:

(1) Responsiveness to rapid changes in market demand, product modification and shorter product life cycles.

(2) High quality products at low cost.

(3) Better use of materials, machinery, personnel and reduced inventory.

(4) Better control of production and management of the total manufacturing operation.

Computer aided design (CAD) allows the designer to conceptualize objects more easily without having to make costly illustrations, models or prototypes. These systems are now capable of rapidly and completely analyzing designs, from a simple bracket to complex structures, such as aircraft wings. Using computer aided engineering (CAE), the performance of structures subjected to static or fluctuating loads and temperatures can now be simulated, analyzed, and tested efficiently, accurately, and more quickly than ever on the computer. The information developed can be stored, retrieved, displayed, printed and transferred anywhere in the organization. Designs can be optimized, and modifications can be made directly and easily at any time.

Computer aided manufacturing (CAM) involves all phases of manufacturing by utilizing and processing further the large amount of information on materials and processes collected and stored in the organization's database. Computers now assist manufacturing engineers and others in organizing tasks such as programming numerical control of machines; programming robots for material handling; designing tools, dies, fixtures and maintaining quality control.

Computer aided process planning (CAPP) is capable of improving productivity in a plant by optimizing process plans, reducing planning costs, and improving the

Chapter 5 Computer-Aided Design and Manufacturing of Materials

consistency of product quality and reliability. Functions such as cost estimating and work standards (time required to perform a certain operation) can also be incorporated into the system.

The high level of automation and flexibility achieved with numerically controlled machines has led to the development of group technology (GT) and cellular manufacturing. The concept of group technology is that parts can be grouped and produced by classifying them according to similarities in design and manufacturing processes. In this way, part designs and process plans can be standardized, and families of parts can be produced efficiently and economically.

Flexible manufacturing systems (FMS) integrate manufacturing cells into a large unit, containing industrial robots serving several machines, all interfaced with a central computer. Flexible manufacturing systems have the highest level of efficiency, sophistication and productivity in manufacturing. They are capable of producing parts randomly and changing manufacturing sequences on different parts quickly. Thus, they can meet rapid changes in market demand for various types of products.

An important concept in manufacturing is just in time production (JIT), which states that supplies are delivered just in time to be used, parts are produced just in time to be made into subassemblies and assemblies, and products are finished just in time to be sold. In this way, inventory carrying costs are low, part defects are detected right away, productivity is increased, and high quality products are made at low cost.

Artificial intelligence (AI) involves the use of machines and computers to replace human intelligence. Computers controlled of learning from experience and making decisions and minimize costs. Expert systems, which are basically intelligent computer programs, are being developed rapidly with capabilities to perform tasks and solve difficult real life problems as well as human experts would. With advances in artificial neural networks, which are being designed to simulate the thought processes of the human brain, these systems have the capability of modeling and simulating production facilities, monitoring and controlling manufacturing processes, diagnosing problems in machine performance, conducting financial planning, and managing a company's manufacturing strategy.

We can now envisage the factory of the future in which production takes place with little or no direct human intervention, The human role will be confined to supervision, maintenance, upgrading of machines, computers and software. The impact of such advances on the workforce remains to be fully assessed.

A word of caution is necessary here because the implementation of some of the modem technologies outlined above requires considerable technical and economic expertise, time, and capital investment. Some of this high technology has been applied improperly, largely because of lack of proper advice and incorrect assessment of the real and specific needs of a company and the market for its products, as well as poor communication among the parties involved such as vendors, suppliers, technical personnel and company management. This technology can be implemented either too soon or on too large or ambitious a scale, involving major expenditures with question able return on investment and disappointing results.

Although large corporations can afford to implement modern technology and take risks, smaller companies generally have difficulty in doing so with their limited personnel, resources and capital. More recently, the concept of shared manufacturing has been proposed. This will consist of a nationwide network of manufacturing facilities, with state of the art equipment for training, prototype development and small scale production runs will be available to help small companies develop products and compete in the international marketplace.

5.1.9 Quality Assurance and Total Quality Management

We all have used terms like poor quality and good quality in describing a product. What do we mean by quality? In a broad sense, quality is a characteristic or property consisting of several well defined technical and aesthetic, hence subjective, considerations. The general public's perception is that a high quality product functions reliably and as expected over a long period of time.

Product quality has always been one of the most important aspects of manufacturing, as it directly influences the marketability of a product and customer satisfaction. Traditionally, quality assurance has been obtained by inspecting parts after they have been manufactured. Parts are inspected to ensure that they conform to a detailed set of specifications and standards such as dimensions, surface finish, and mechanical and physical properties.

However, quality cannot be inspected into a product after it is made. The practice of inspecting products after they are made is now being replaced rapidly by the broader view that quality must be built into a product from the design stage through all subsequent stages of manufacture and assembly. Because products are made by using several manufacturing processes which can have significant variations in their performance even within a short period of time, the control of processes is a critical factor in product quality. Thus, we control processes and not products.

Chapter 5 Computer-Aided Design and Manufacturing of Materials

Producing defective products can be very costly to the manufacturer, creating difficulties in assembly operations, necessitating repairs in the field, and resulting in customer dissatisfaction. Contrary to general public opinion, low quality products do not necessarily cost less than high quality products.

Although it can be described in various terms, product integrity is a term that can be used to define the degree to which a product is suitable for its intended purpose; fills a real market need; functions reliably during its life expectancy; can be maintained with relative ease. Product integrity has also been defined as the total product experience of the customer, or as the totality of qualities needed to conceive, produce, and market the product successfully.

Total quality management (TQM) and quality assurance have become the responsibility of everyone involved in designing and manufacturing a product. Our awareness of the technological and economic importance of built in product quality has been heightened further by recent pioneers in quality control, primarily Deming, Taguchi and Juran. They pointed out the importance of management's commitment to product quality, pride of workmanship at all levels of production, and the use of powerful techniques such as statistical process control (SPC) and control charts for on line monitoring of part production and rapidly identifying sources of quality problems. Ultimately, the major goal is to prevent defects from occurring, rather than detect defective products. We can now produce computer chips in which only a few parts out of a million may be defective. The importance of this trend in the reliability of products and customer satisfaction is self evident.

Important developments in quality assurance include the implementation of experimental design, a technique in which the factors involved in a manufacturing, process and their interactions are studied simultaneously. Thus, for example, variables affecting dimensional accuracy or surface finish in a machining operation can be readily identified, allowing appropriate actions to be taken. The use of computers has greatly enhanced our capability to utilize such techniques rapidly and effectively.

The strong trend in global manufacturing and competitiveness has created the need for international conformity and consensus regarding the establishment of quality control methods. This has resulted in the International Organization for Standardization ISO 9000 series on Quality Management and Quality Assurance Standards. A company's registration for this standard, which is a quality process certification and not a product certification, means the company conforms to consistent practices as specified by its own quality system. ISO 9000 has permanently influenced the manner in which companies conduct business in world trade and is rapidly becoming the world standard for quality.

5.1.10　Organization for Manufacture

The various manufacturing activities and functions that we have described must be organized and managed efficiently and effectively in order to maximize productivity and minimize costs, while maintaining high quality standards. Because of the complex interactions among the various factors involved in manufacturing materials, machines, people, information, power, and capital the proper coordination and administration of diverse functions and responsibilities are essential.

Manufacturing engineers traditionally have several major responsibilities:

(1) Plan the manufacture of a product and the processes to be utilized. This function requires a thorough knowledge of the product, its expected performance and specifications.

(2) Identify machines, equipment, tooling and the personnel to carry out the plan. This function requires evaluation of the capabilities of machines, tools and workers, so that proper functions and responsibilities can be assigned.

(3) Interact with design and materials engineers to optimize productivity and minimize production costs.

(4) Cooperate with industrial engineers when planning plant floor activities such as plant layout, machine arrangement, material handling equipment, time and motion study, production methods analysis, production planning and scheduling, and maintenance. Some of these activities are carried out under the plant engineering, and some are interchangeably performed by both manufacturing and industrial engineers.

Manufacturing engineers, in cooperation with industrial engineers, also are responsible for evaluating new technologies and their applications, and how they can be implemented. In view of the rapidly growing amount of technical information, this task in itself can present a major challenge. Gaining a broad perspective of computer capabilities, applications and integration in all phases of manufacturing operations is important. This knowledge is particularly crucial for long range production and facility planning in view of constantly changing market demand and product mix.

In order to respond to these major changes, it is essential in a manufacturing organization to:

(1) View the people in the organization as important assets.

Chapter 5 Computer-Aided Design and Manufacturing of Materials

(2) Emphasize the importance and need for teamwork and involvement in problem solving and in decision making processes in all aspects of operations.

(3) Encourage product innovation and improvements in productivity.

(4) Encourage efforts for continuous improvement in quality (quality first).

(5) Increase flexibility of operation for faster response to product demands in both the domestic and the global marketplace (economy of time).

(6) Ultimately and most importantly, focus on customer satisfaction.

5.2 Definition of CAD/CAM

Computer-aided design (CAD) can be defined as the use of computer systems to assist in the creation, modification, analysis or optimization of a design. Computer-aided manufacturing (CAM) can be defined as the use of computer systems to plan, manage and control the operations of a manufacturing plant through either direct or indirect computer interface with the plant's production resources. CAD/CAM is considered to be a multi-disciplines, knowledge-intensive[2] and high and new technology.

CAD/CAM is a term which means computer-aided design and computer-aided manufacturing. It is the technology concerned with the use of digital computers to perform certain functions in design and production. This technology is moving in the direction of greater integration of design and manufacturing, two activities which have traditionally been treated as distinct and separate functions in a production firm. Ultimately, CAD/CAM will provide the technology base for the computer-integrated factory of the future.

Computer-aided design (CAD) can be defined as the use of computer systems to assist in the creation, modification, analysis or optimization of a design. The computer systems consist of the hardware and software to perform the specialized design functions required by the particular user firm. The CAD hardware typically includes the computer, one or more graphics display terminals, keyboards and other peripheral equipment. The CAD software consists of the computer programs to implement computer graphics on the system plus application programs to facilitate the engineering functions of the user company. Examples of these application programs include stress-strain analysis of components, dynamic response of mechanisms, heat-transfer calculations[3] and numerical control part programming. The collection of application programs will vary from one user firm to the next because their product lines, manufacturing processes and customer markets are different. These factors give rise to differences in CAD system requirements.

Computer-aided manufacturing (CAM) can be defined as the use of computer systems to plan, manage and control the operations of a manufacturing plant through either direct or indirect computer interface with the plant's production resources. As indicated by the definition, the applications of computer aided manufacturing fall into two broad categories:

(1) Computer monitoring and control

These are the direct applications in which the computer is connected directly to the manufacturing process for the purpose of monitoring or controlling the process.

(2) Manufacturing support applications

These are the indirect applications in which the computer is used in support of the production operations in the plant, but there is no direct interface between the computer and the manufacturing process.

The distinction between the two categories is fundamental to an understanding of computer-aided manufacturing. It seems appropriate to elaborate on our brief definitions of the two types.

Computer monitoring and control can be separated into monitoring applications and control applications. Computer process monitoring involves a direct computer interface with the manufacturing process for the purpose of observing the process and associated equipment and collecting data from the process. The computer is not used to control the operation directly. The control of the process remains in the hands of human operators, who may be guided by the information compiled by the computer.

Computer process control goes one step further than monitoring by not only observing the process but also controlling it based on the observations. The distinction between monitoring and control is displayed in Figure 5.3. With computer monitoring the flow of data between the process and the computer is in one direction only, from the process to the computer. In control, the computer interface allows for a two-way flow of data. Signals are transmitted from the process to the computer just as in the case of computer monitoring. In addition, the computer issues command signals directly to the manufacturing process based on control algorithms contained in its software.

Figure 5.3 Computer monitoring (a) versus computer control (b)

In addition to the applications involving a direct computer-process interface for

Chapter 5 Computer-Aided Design and Manufacturing of Materials

the purpose of process monitoring and control, computer-aided manufacturing also includes in direct applications in which the computer serves a support role in the manufacturing operations of the plant. In these applications, the computer is not linked directly to the manufacturing process. Instead, the computer is used "off-line[④]" to provide plans, schedules, forecasts, instructions and information by which the firm's production resources can be managed more effectively. The form of the relationship between the computer and the process is represented symbolically in Figure 5.4.

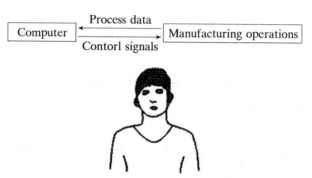

Figure 5.4 CAM for manufacturing support

Dashed lines are used to indicate that the communication and control link is an off-line connection, with human beings often required to consummate the interface. Some examples of CAM for manufacturing support that are discussed in subsequent chapters of this book include:

(1) Numerical control part programming by computers

Control programs are prepared for automated machine tools.

(2) Computer-automated process planning

The computer prepares a listing of the operation sequence required to process a particular product or component.

(3) Computer-generated work standards

The computer determines the time standard for a particular production operation.

(4) Production scheduling

The computer determines an appropriate schedule for meeting production requirements.

(5) Material requirements planning

The computer is used to determine when to order raw materials and purchase components and how many should be ordered to achieve the production schedule.

(6) Shop floor control

In this CAM application, data are collected from the factory to determine progress of the various production shop orders.

In all of these examples, human beings are presently required in the application either to provide input to the computer programs or to interpret the computer output and implement the required action.

5.3 Rational for CAD/CAM

The rational for CAD/CAM is similar to that used to justify any technology-based improvement in manufacturing. It grows out of a need to continually improve productivity, quality, and, in turn, competitiveness. There are also other reasons why a company might make a conversion from process to CAD/CAM: increased productivity; better quality; better communication; common database with manufacturing; reduced prototype construction costs; faster response to customers

(1) Increased productivity

Productivity in the design process is increased by CAD/CAM. Time-consuming tasks such as mathematical calculations, data storage and design visualization are handled by the computer, which gives the designer more time to spend on conceptualizing and completing the design. In addition, the amount of time required to document a design can be reduced significantly with CAD/CAM. All of these taken together means a shorter design cycle, shorter overall project completion time and a higher level of productivity.

(2) Better quality

Because CAD/CAM allows designers to focus more on actual design problems and less on time-consuming, nonproductive tasks, product quality improves with CAD/CAM. CAD/CAM allows designers to examine a wider range of design alternatives (e. g., product features) and to analyze each alternative more thoroughly before selecting one. In addition, because labor-intensive[⑤] tasks are performed by the computer, fewer design errors occur. These all lead to better quality.

(3) Better communication

Design documents such as drawings, part lists, bills of material and specifications are tools used to communicate the design to those who will manufacture it. The more uniform, standardized and accurate these tools are, the better the communication will be. Because CAD/CAM leads to more uniform, standardized and accurate documentation, it improves communication.

Chapter 5 Computer-Aided Design and Manufacturing of Materials

(4) Common database with manufacturing

This is one of the most important benefits of CAD/CAM. With CAD/CAM, the data generated during the design of a product can be used in producing the product. This sharing of a common database helps to eliminate the age-old^⑥ "wall" separating the design and manufacturing functions.

(5) Reduced prototype construction costs

With manual design, models and prototypes of a design must be made and tested, adding to the cost of the finished product. With CAD/CAM, 3D computer models can reduce and, in some cases, eliminate the need for building expensive prototypes, Such CAD/CAM capabilities as solids modeling allow designers to substitute computer models for prototype in many cases.

(6) Faster response to customers

To customers response time is critical in manufacturing. How long does it take to fill a customer's order? The shorter the time, the better it is. A fast response time is one of the keys to being more competitive in an increasingly competitive marketplace. Today, the manufacturer with the faster response time is as likely to win a contract as the one with the lowest bid. By shortening the overall design cycle and improving communication between the design and manufacturing component, CAD/CAM can improve a company's response time.

5.4 Numerical Control and Numerical Control Machine

5.4.1 Numerical Control

Controlling a machine tool using a punched tape or stored program is known as numerical control (NC). NC has been defined by the Electronic Industries Association (EIA) as "a system in which actions are controlled by the direct insertion of numerical data at some point. The system must automatically interpret at least some portion of this data." The numerical data required to produce a part is known as a part program.

A numerical control machine tool system contains a machine control unit (MCU) and the machine tool itself. The MCU is further divided into two elements: The data processing unit (DPU) and the control loops unit (CLU). The DPU processes the coded data from the tape or other media and passes information on the position of each axis, required direction of motion, feed rate, and auxiliary function control signals to the CLU. The CLU operates the drive mechanisms of the machine, receives feed back signals concerning the actual position and velocity of each of the axes, and signals the completion of operation. The DPU sequentially reads the data. When each line has completed execution as noted by the CLU, another line of data is read. Geometric and kinematic data are typically fed from the DPU to the CLU and CLU then governs the physical system based on the data from the DPU.

Numerical control was developed to overcome the limitation of human operators, and it has done so. Numerical control machines are more accurate than manually operated machines, they can produce parts more uniformly, they are faster, and the long-run[①] tooling costs are lower. The development of NC led to the development of several other innovations in manufacturing technology:

(1) Electric discharge machining.

(2) Laser-cutting.

(3) Electron beam welding.

Numerical control has also made machine tools more versatile than their manually operated predecessors. An NC machine tool can automatically produce a wide variety of parts, each involving an assortment of widely varied and complex machining processes. Numerical control has allowed manufacturers to undertake the production of products that would not have been feasible from an economic perspective using manually controlled machine tools and processes.

5.4.2 Classifications of Numerical Control Machines

Numerical control machines are classified in different ways. An early method was to categorize them as being either point-to-point or continuous-path[②] machines. Point-to-point machines, as the name implies, move in a series of steps from one point to the next (Figure 5.5). Point-to-point machines are less sophisticated and less precise than continuous-path machines. Continuous-path machines move uniformly and evenly along the cutting path rather than through a series of horizontal and vertical steps. Such machines are more sophisticated and require more memory in the MCU than point-to-point machines. Figure 5.6 illustrates the types of cutting paths performed by continuous-path machines.

Chapter 5　Computer-Aided Design and Manufacturing of Materials

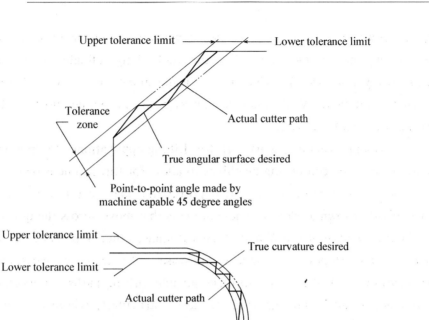

Figure 5.5　Point-to-point path

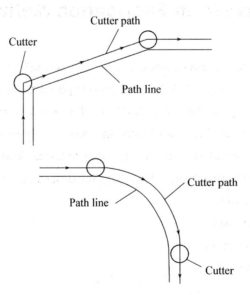

Figure 5.6　Continuous path

Another way to classify NC machines is as either positioning or contouring machines. Point-to-point machines are considered positioning machines. Continuous-path machines are considered contouring machines. Positioning machines have as few as two axes: the X axis and the Y axis. Contouring machines must have at least three axes: the X, Y and Z axes.

Note that X represents the longitudinal axes, Y the transverse axis and Z the up-and-down or vertical axis. On some machines, movement is accomplished by positioning the spindle, and thus the tool, longitudinally along the X axis, transversely along the Y axis, and vertically along the Z axis. The workpiece is affixed to the table. With other machines, both the spindle and the table (thus the workpiece) can be moved.

Positioning machines work well for drilling applications. Milling operations are more likely to be contouring machines to allow for three-dimensional control.

Some of the more sophisticated positioning machines are able to accomplish angular cuts known as slopes. These are cuts that move across the quadrants formed by the intersection of the X and Y axes at angles other than 90 degree to either the X or Y axis. Slopes are generally imprecise and inaccurate. However, there are instances in which the ability to make angular cutting paths is important. In these cases, slopes can be an important feature, particularly where the cut surfaces do have to mate with another surface. When precise, accurate angular cutting paths must be made, a contouring machine is needed.

5.5 Solid Freeform Fabrication Methods

Several manufacturing processes are available to make the important transition from computer aided design (CAD) to a prototype part. Several new technologies began to make their appearance after 1987. In that year, stereolithography was first introduced by 3D Systems Inc., and over the next five years several rival methods also appeared. This created the family of processes known as solid freeform fabrication (SFF). SFF processes are sometimes described as:

(1) Parts on demand.
(2) From part to part.
(3) Desktop manufacturing.
(4) Rapid prototyping.

5.5.1 Stereolithography

Stereolithography apparatus (SLA) was invented by Charle Hull of 3D Systems Inc. It is the first commercially available rapid prototyper and is considered as the most widely used prototyping machine. The material used is liquid photo-curable[9] resin, acryrlate. Under the initiation of photons, small molecules (monomers) are polymerized into large molecules. Based on this principle, the part is built in a vat of liquid resin.

The SLA machine creates the prototype by tracing layer cross-sections on the surface of the liquid photopolymer pool with a laser beam. Unlike the contouring or zigzag cutter movement used in CNC machining, the beam traces in parallel lines, or vectorizing first in one direction and then in the orthogonal direction. An elevator table in the resin vat rests just below the liquid surface whose depth is the light absorption limit. The laser beam is deflected horizontally in X and Y axes by galvanometer-driven mirrors so that it moves across the surface of the resin to produce a solid pattern. After a layer is built, the elevator drops a user-specified[①] distance and a new coating of liquid resin covers the solidified layer. A wiper helps spread the viscous polymer over for building the next layer. The laser draws a new layer on the top of the previous one. In this way, the model is built layer by layer from bottom to top. When all layers are completed, the prototype is about 95 % cured. Post-curing is needed to completely solidify the prototype. This is done in a fluorescent oven where ultraviolet light floods the object (prototype). There are several features worthy of mentioning of SLA.

(1) Material

There are five commercially available photopolymers. All of them are a kind of acrylate.

(2) Support

Because a model is created in liquid, the overhanging regions of the part (unsupported below) may sag or float away during the building process. The prototype thus needs some predesigned support, until it is cured or solidified. The support can be pillars, bridges and trusses. Sometimes posts or internal honeycomb sections are needed to add rigidity to tall thin-walled shapes during the process. These additional features are built on the model parts and have to be trimmed after the model building is completed.

(3) Model accuracy and performance

The accuracy achieved is about 0.1 % of the overall dimension and deteriorates with larger sizes but no more than 0.5 %. The layer thickness is between 0.004 in and 0.03 in (1 in = 25.4 mm).

The photopolymer-made prototype is brittle and may not be strong enough to withstand high stress testing. Also, the shrinkage of the material may make the prototype deform.

(4) Capacity

The size of the vat that holds the liquid polymer determines the size of the prototype that can be built.

(5) Recycling

Photopolymers are thermoset material and cannot be melted again for reuse.

5.5.2 Selective Laser Sintering

Another very popular method is selective laser sintering (SLS), commercialized by the DTM Corporation. In many respects SLS is similar to SLA except that the laser is used to sinter and fuse powder rather than photocure a polymeric liquid.

The first step is to prepare the "STL/SLI" files as described earlier. Inside the SLS machine, a thin layer of fusible powder is laid down and heated to just below its melting point by infrared heating panels at the side of the chamber. Then a laser sinters and fuses the desired pattern of the first slice of the object in the powder. Next, this first fused slice descends, the roller spreads out another layer of powder, and the process repeats.

In comparison with SLA, this process can rely on the supporting strength of the unfused powder around the partially fused object. Therefore, support columns for any overhanging parts of the component are not needed. This allows the creation of rather delicate, lacelike objects. Nevertheless hand finishing is still needed to improve the inevitable stair-stepping. Also, SLS parts have a rough, grainy appearance from the sintering process, and it is often preferable to hand smooth the surfaces. Another difficulty is maintaining the temperature of the powder at a few degrees below melting. This is done with the infrared panels, but maintaining an even temperature over a large mass of powder requires long periods of stabilization before sintering by the laser can be started.

5.5.3 Laminated Object Modeling

Laminated object modeling (LOM) was developed by Helisys Inc., and like SLA and SLS, it was first offered commercially in the period from 1987 to 1990. In LOM, the laser is used to cut the top slice of a stack of paper that is progressively glued together. After each profile has been cut by the laser (Figure 5.7), the roll of paper is advanced, a new layer is glued onto the stack, and the process is repeated. After fabrication, some trimming, hand finishing and curing are needed. For larger components, especially in the automobile industry, LOM is often preferred over the SLA or SLS processes.

Chapter 5 Computer-Aided Design and Manufacturing of Materials

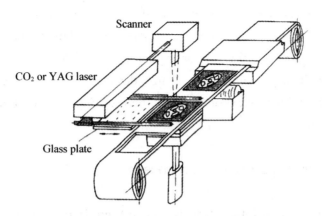

Figure 5.7 Laminated object modeling

5.5.4 Fused Deposition Modeling

Fused deposition modeling (FDM) was developed by Stratasys Inc. Figure 5.8 shows that the material is supplied as a filament from spool. The overall geometry and system are reminiscent of icing a cake. The filament melts as it flows through a heated delivery head and emerges as a thin ribbon through an exit nozzle. The nozzle is guided around by CNC code, and the viscous ribbon of polymer is gradually built up from a fixtureless base plate. In terms of motion control, FDM is more similar to CNC machining than SLA or SLS. For simple parts, there is no need for

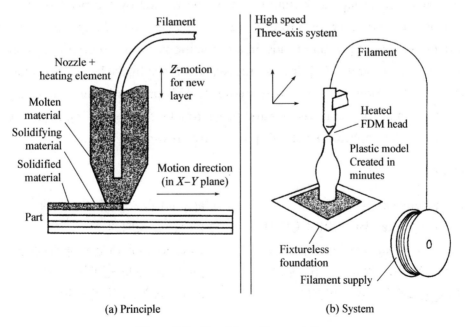

(a) Principle   (b) System

Figure 5.8 Fused deposition modeling

fixturing, and material can be built up layer by layer. The creation of more complex parts with inner cavities, unusual sculptured surfaces, and overhanging features does require a support base, but the supporting material can be broken away by hand, thus requiring minimal finishing work. Thus, despite the similarities with the CNC machine from the point of view of control, the resulting parts that can be made are more in the SFF family.

5.5.5 3D Plotting and Printing Processes

Several types of 3D printing processes have been developed in recent years and are constantly being updated at the time of this writing. Some of these are aimed at the educational CAD/CAM market where students are invited to obtain quick models of an emerging design. At the same time, such machines might be useful in an industrial design studio, where artists might want to generate and regenerate a quick succession of prototypes for the "look and feel" of an emerging design. Examples include:

(1) 3D printing of cornstarch, followed by layer-by-layer binder hardening, is the basic principle behind the Z-corporation machine. The first step is to spread a thin layer of powder of the desired material across the top of the bed. The next step hardens the desired geometry into this layer of powder. The hardening is not done with a laser (like SLS) but with a binder phase. Fine droplets of the binder stream are printed down through a continuous-jet nozzle carried by the print head. Since material is built up layer by layer in an x/y plane, the process resembles the motions of the ink-jet printing heads on a conventional word processing printer.

(2) A more accurate 3D printing process, developed by Sachs and his colleagues at MIT, was the forerunner to this technology. This process is being used to build the ceramic molds for metal castings and powder-metal tooling for injection molding dies. Commercial applications of this process are growing.

Notes

① hole generation 空穴产生
② knowledge-intensive 知识密集型的
③ heat-transfer calculation 传热计算
④ off-line 脱机的;离线的,未连线的
⑤ labor-intensive 劳动密集型的
⑥ age-old 古老的;由来已久的
⑦ long-run 长期的
⑧ continuous-path 连续通路;连续路径
⑨ photo-curable 光固化的
⑩ user-specified 使用者指定的

Chapter 5 Computer-Aided Design and Manufacturing of Materials

Vocabulary

acryrlate [ˈækrɪleɪt] n. 丙烯酸盐；丙烯酸酯

aesthetic [iːsˈθetɪk] adj. 美的；美学的；审美的

backbone [ˈbækbəʊn] n. 支柱

beverage [ˈbev(ə)rɪdʒ] n. 饮料

cellular [ˈseljʊlə] n. 移动电话；单元 adj. 细胞的；多孔的；由细胞组成的

chamber [ˈtʃeɪmbə] n. （身体或器官内的）室，腔；房间；会所

competitive [kəmˈpetɪtɪv] adj. 竞争的；比赛的；求胜心切的

complexity [kəmˈpleksətɪ] n. 复杂，复杂性；复杂错综的事物

conceptual [kənˈseptjʊəl] adj. 概念上的

concurrent [kənˈkʌr(ə)nt] n. [数]共点；同时发生的事件 adj. 并发的；一致的；同时发生的

controversial [kɒntrəˈvɜːʃ(ə)l] adj. 有争议的；有争论的

documentation [ˌdɒkjʊmenˈteɪʃ(ə)n] n. 文件，文件编制，文档

elastomer [ɪˈlæstəmə] n. [力]弹性体，[高分子]高弹体

envisage [ɪnˈvɪzɪdʒ] vt. 正视，面对；想像

expendable [ɪkˈspendəb(ə)l] n. 消耗品 adj. 可消费的

fluorescent [flʊəˈres(ə)nt] n. 荧光；日光灯 adj. 荧光的；萤光的；发亮的

galvanometer [ˌgælvəˈnɒmɪtə] n. 检流计，[电]电流计

geometric [dʒɪəˈmetrɪk] adj. 几何学的；[数]几何学图形的

honeycomb [ˈhʌnɪkəʊm] n. 蜂巢，蜂巢状之物

infrequent [ɪnˈfriːkw(ə)nt] adj. 罕见的；稀少的；珍贵的；不频发的

integration [ˌɪntɪˈgreɪʃ(ə)n] n. 集成；综合

integrity [ɪnˈtegrɪtɪ] n. 诚信；正直；完整性

interchangeably [ˌɪntəˈtʃeɪndʒəblɪ] adv. [数]可交换地

intervention [ˌɪntəˈvenʃ(ə)n] n. 介入；调停；妨碍

intuition [ˌɪntjʊˈɪʃ(ə)n] n. 直觉；直觉力；直觉的知识

inventory [ˈɪnv(ə)nt(ə)rɪ] n. 存货，存货清单；详细目录；财产清册

iterative [ˈɪt(ə)rətɪv] adj. [数]迭代的；重复的，反复的 n. 反复体

kinematic [ˌkɪnɪˈmætɪk] adj. [力]运动学上的，[力]运动学的

lawnmower [ˈlɔːnməʊə] n. [建]剪草机

machinery [məˈʃiːn(ə)rɪ] n. 机械；机器；机构；机械装置

marketability [ˌmɑːkɪtəˈbɪlɪtɪ] n. 可销售，市场性；适销性

methodology [ˌmeθəˈdɒlədʒɪ] n. 方法学，方法论

monetary [ˈmʌnɪt(ə)rɪ] adj. 货币的；财政的

multifunctional [ˌmʌltɪˈfʌŋkʃənəl] adj. 多功能的

nonproductive [ˌnɒnprəˈdʌktɪv] adj. 非生产性的；对生产无直接关系的

obsolescence [ˌɒbsəˈlesəns] n. [生物]退化；荒废

optimum [ˈɒptɪməm] n. 最佳效果；最适

Chapter 5　Computer-Aided Design and Manufacturing of Materials

宜条件　*adj*. 最适宜的
overdesign [ˌəʊvədɪˈzaɪn] *n*. 保险设计；超安全标准的设计　*vt*. 保险设计；超安全标准设计
permanent [ˈpɜːm(ə)nənt] *n*. 烫发（等于 permanent wave）　*adj*. 永久的，永恒的；不变的
photocure [fəʊtəʊkˈjʊə] *n*. 光固化
producibility [prəʊˌdjuːsəˈbɪlətɪ] *n*. 可生产性；可生产
productivity [prɒdʌkˈtɪvɪtɪ] *n*. 生产力；生产率；生产能力
prototype [ˈprəʊtətaɪp] *n*. 原型；标准，模范
rational [ˈræʃ(ə)n(ə)l] *n*. 有理数　*adj*. 合理的；理性的
repetitive [rɪˈpetɪtɪv] *adj*. 重复的
resin [ˈrezɪn] *n*. 树脂；松香　*vt*. 涂树脂；用树脂处理
sculptured [ˈskʌlptʃəd] *adj*. 用刻纹装饰的；具刻纹的　*v*. 雕刻；用雕塑装饰（sculpture 的过去式）
sophistication [səˌfɪstɪˈkeɪʃn] *n*. 复杂；诡辩；老于世故；有教养
spool [spuːl] *n*. 线轴；缠线框　*vt*. 缠绕；卷在线轴上
standardization [ˌstændədaɪˈzeɪʃn] *n*. 标准化；[数]规格化；校准
stapler [ˈsteɪplə(r)] *n*. 订书机，订书机，订书器
stereolithography [steˌrɪəlɪˈθɒɡrəfɪ] *n*. 光固化
thermoset [ˈθɜːməʊset] *n*. 热固性；热凝物　*adj*. 热固性的
toaster [ˈtəʊstə] *n*. 烤面包器，烤箱；烤面包的人
ultraprecision [ˌʌltreɪprɪˈsɪʒn] *n*. 超精度
vat [væt] *n*. 大桶；瓮染料制剂桶　*vt*. 把……盛入大桶；把……放入大桶里染

◆ **Exercises**

1. Translate the following Chinese phrases into English

(1) 计算机辅助设计　　(2) 计算机集成制造　　(3) 同步工程
(4) 设计优化　　　　　(5) 制造和装配　　　　(6) 黑色金属
(7) 低熔点合金　　　　(8) 表面处理　　　　　(9) 涂层和镀层
(10) 计算机数控　　　 (11) 表面光洁度　　　 (12) 3D 计算机模型
(13) 激光切割　　　　 (14) 点对点　　　　　 (15) 辅助硬化
(16) 薄壁形状　　　　 (17) 熔融沉积成型

2. Translate the following English phrases into Chinese

(1) computer aided manufacturing　　(2) concurrent engineering
(3) design and manufacturing　　　　(4) conceptual designs
(5) nonferrous metals and alloys　　(6) refractory metals
(7) precious metals　　　　　　　　 (8) expendable mold
(9) machine control system　　　　　(10) adaptive control
(11) quality management　　　　　　 (12) numerical control
(13) electron beam welding　　　　　(14) three-dimensional control

(15) solid freeform fabrication (16) selective laser sintering

3. Translate the following Chinese sentences into English

(1) 至于生产的原材料如何选择,我们还需要了解不同材料的特性。材料的特性被归为力学性能、物理性能、热性能、化学性能、电学性能、光学性能和声学性能等几类。

(2) 制造工艺可分为两类:初级加工和次级加工。初级加工就是将木材、金属等原材料变为标准坯料,之后标准坯料需要进一步加工才能成为最终产品,这一过程称为次级加工。

(3) 用电子图板进行二维绘图设计是实现计算机辅助设计(CAD)与计算机辅助制造(CAM)一体化的基础。

(4) 计算机应用领域包括科学计算、数据处理和信息管理、自动控制、辅助设计制造和测试、系统仿真。

(5) 这些模块执行具体的业务任务,如资本投资管理、人力资源管理和质量管理等。

(6) 概念设计方案的评价方法是设计过程可接受性决策的重要依据。

(7) 物以稀为贵,钨,是一种宝贵的具有战略意义的稀有金属。

(8) 从成本上看,这个例子清楚地表明,用点对点集成架构建立一个大规模 HIN 是完全不可行的。

4. Translate the following English sentences into Chinese

(1) Most products require a surface finish which provides protection and makes them more attractive. This final process is called finishing. One type of finishing involves coating over the surface of the material. The other is a conversion of the surface itself. Finishing makes the products both more attractive and durable.

(2) Floating point solutions offers a wide range of products in the fields of computer aided manufacturing (CAM), reverse engineering (RE) and rapid prototyping (RP).

(3) Computer aided process planning (CAPP) acts as the bridge between computer aided design and computer aided manufacturing, which has an important effect on quality and cost of product development.

(4) CNC (Computer numerical control) is a technology originally developed for the aviation industry, enabling the precise construction of digitally generated, complex curved objects.

(5) Since numerical control was adopted machine tools, the productivity has been raised greatly.

(6) In reasoning process of product conceptual design, expression of knowledge of design problem is a pivotal step.

5. Translate the following Chinese essay into English

数控机床最基本的功能是自动、精确、连续的运动控制。一切形式的数控设备都有两个或多个运动方向,称为轴。这些轴可以沿其行程精确自动定位。最常见的两种类型的轴是线性轴(沿直线轨迹运动)和旋转轴(沿环形轨迹运动)。

与传统机床的运动需要人工转动曲柄和手轮不同,数控机床由数控下的伺服电机驱动,依照程序运动。一般而言,几乎所有的数控机床都能够对运动类型(快速、线形和圆形)、

运动轴、运动量和运动速度(进给速度)进行编程。

6. Translate the following English essay into Chinese

Computer aided engineering (CAE) is the use of computer to support engineers in tasks such as design, analysis, simulation, manufacture, planning, diagnosis and repair. CAE tools are very widely used in the automotive industry. In fact, their use has enabled the automakers to reduce product development cost and time while improving the safety, comfort and durability of the vehicles they produce.

The predictive capability of CAE tools has progressed to the point where much of the design verification is now done using computer simulations rather than physical prototype testing. CAE dependability is based upon all proper assumptions as inputs and must identify critical inputs.

Even though there have been many advances in CAE and it is widely used in the engineering field, physical testing is still used as a final confirmation for subsystems due to the fact that CAE cannot predict all variables in complex assemblies.

CAE applications: (1) Stress analysis on components and assembliesusing finite element analysis (FEA); (2) Thermal and fluid flow analysis; (3) Kinematics; (4) Mechanical event simulation (MES); (5) Analysis tools for process simulation of operations such as casting, molding and die press forming; (6) Optimization of the product or process.

扫一扫,查看更多资料

Part III
Production and Processing of Materials

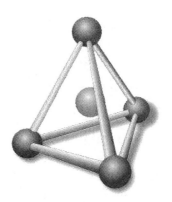

Part II
Production and Processing of Materials

Chapter 6 Powder Metallurgy

6.1 Introduction

The forming or shaping of materials can be done from the solid, granular, or liquid state-referring to the state of the work material in the shaping phase prior to stabilization of the material. Shaping and stabilization can sometimes be integrated.

The manufacture of a product from the granular or particle state covers, in general, a broad spectrum of materials and components or products, such as carbide tools (sintered or cemented tool inserts), metal powder components, sand molds, ceramics, concrete, tablets and bread.

A granular material is a mixture of solid grains or particles possibly of varying sizes. Each grain or particle may be a combination of smaller units, for example, the crystals in metal grains. The use of granular materials is generally due to one or more of the following reasons.

(1) The particular material is only available or can only be produced in the granular state.

(2) The desired properties (porosity, combination of materials, *etc.*) can only be obtained from granular materials.

(3) Manufacture of the product is cheaper than by other methods.

(4) Small components are difficult to produce by other methods.

Production of components from granular materials will generally follow the same pattern:

(1) Production of the granular material.

(2) Conditioning or preparation for shaping and stabilization.

(3) Shaping.

(4) Stabilization of the shape.

(5) Finishing operations.

Depending on the material and the requirements of the component, these phases involve different basic processes. In this context, only the production of metal powder components will be discussed; this area is, in general, called powder metallurgy.

Within the last decade, the production of components from metal powders has increased rapidly, with a probable yearly expansion of 10~20 % in the years to come. This is due to one or more of the following reasons:

(1) The production of simple or complicated geometries can be performed in one operation and with high dimensional accuracy.

(2) A high (nearly 100 %) material utilization.

(3) The final properties, even if they are not on the same level as corresponding solid materials, are satisfactory for most applications.

(4) The production of components that can be produced by other methods is only with difficulty.

(5) Powder metallurgy is the most economic process in many situations.

Powder metallurgy competes primarily with casting, hot and cold forging, and cutting. These methods normally involve several operations. In the following sections, the characteristics of powder metallurgical processes (powder, preparation, compacting, sintering, post-sintering treatments, *etc.*), properties and applications are discussed.

6.2 Characteristics of Powder Metallurgical Processes

Production of a component by powder metallurgy techniques normally involves the following stages or phases.

(1) Production of selection of powder.

(2) Preparation, including the mixing and blending.

(3) Pressing or compacting.

(4) Sintering or heat treatment.

(5) Post-sintering treatment, if necessary.

6.2.1 Metal Powders

Different methods for the production of metal powders have been developed. The most important are the reduction of ores, atomizing and electrolytic deposition. In conventional powder metallurgy, powders produced by the reduction of ores are used extensively, but in recent years the application of powders produced by atomization has increased rapidly. Powders produced by electrolytic deposition are used only for special purposes, and the market for these powders is decreasing. The types and properties of the powders have a major influence on the final properties of the component, so that a fundamental knowledge of powders is important.

6.2.2 Powders Produced by Reduction of Ores

The reduction process is used primarily to produce iron powders. The purity of the powder is directly related to the purity of the ore. In the production of iron powders, pure and dry ore is heat-treated in sealed drums together with coal dust, coke, gravel and chalk at about 1 200 ℃ for about 90 h (the Hoeganaes method[①]). After this reduction, the resulting iron sponge cake is crushed, ground, and heat-treated in a hydrogen atmosphere to provide a reduction of the oxides and to anneal the powder particles. The powders contain the impurities from the ores, and a single grain has many internal pores, making the powder unsuitable for pressing to very high densities, as it will require enormously high pressure to close these internal pores. Depending on the production conditions, the number and size of the internal pores vary, and the general shape of the grains is irregular. In addition to iron powder, nonferrous metal, cobalt, molybdenum and tungsten powders can be produced by the reduction of ores.

6.2.3 Powders Produced by Atomization

In atomization, the powders are produced from the liquid state, which gives great freedom in the choice of materials and in the alloying process. The purity of the powders is directly related to the raw materials and the melting and refining processes. The shapes and sizes of the particles can be varied within wide limits, depending on process parameters.

A flow of liquid metal through an orifice is broken up by a jet stream of gas (air or inert gases), water steam or water. Gas atomization gives spherical and large particles; water atomization gives smaller and irregular grains without internal pores.

Atomization can be used to produce powders of iron, steels (including stainless), lead, zinc, aluminum, bronzes, brasses, and so on.

The use of powders produced by atomization has increased rapidly due to the purity obtainable, the alloying possibilities and the powder properties. It should be mentioned that in the past, the price for atomization powders has been higher than for reduction powders, but now these prices are comparable.

6.2.4 Powders Produced by Electrolytic Deposition

After electrolytic deposition, the metal is crushed and ground by mill grinding to the desired grain sizes. Iron powders produced by electrolytic deposition are more expensive than those produced by reduction or atomization. The electrolytic powders are used only where their special properties (including high purity, density and compressibility) can be utilized.

As mentioned previously, several other powder manufacturing processes exist. They will not be described here, as the powders produced by these processes are used only for special applications.

A metal powder can be characterized by: chemical composition, particle-size distribution, particle shape (spherical, irregular), surface characteristics, internal structure (pores, *etc.*), flow ability (ASTM 213.48/212.48), compressibility, green strength (strength after compaction), sintering properties or sintering abilities (change of dimensions, strengths, *etc.*).

To describe these characteristics, several testing methods have been developed (SAE/ASTM/MPI standards or recommendations).

Powder manufacturers supply all the necessary information about their powders, and these should be studied carefully before selecting a powder for a specific application. Most of the listed characteristics or properties of the powder influence the pressing and sintering processes as well as the green and final strengths.

6.2.5 Preparation of the Powder

An important stage in the production of metal powder components is the preparation of the powder for compaction and sintering. The preparation of a powder consists mainly of mixing or blending to obtain a uniform distribution of the different particle sizes, a coating of the particles with a lubricant, and a uniform distribution of the base powder and alloying elements. The mixing process must be carried out carefully. Too heavy mixing may cause strain hardening, wear of the particles against each other, layering, and so on. Recommendations from the manufacturer must be followed.

Lubrication can be provided as internal or external lubrication. In internal lubrication, a lubricant (zinc stearate or stearic acid, 0.25~1% by weight) is mixed with the powder, increasing its compressibility and decreasing its green strength. After pressing, the lubricant is driven out by heat treatment (in air at 375~425 ℃), before sintering in a controlled atmosphere. In external lubrication, only the die walls are lubricated, avoiding the heat treatment necessary to drive out the lubricant, but this method does not provide the improved flow and compaction properties.

Considering the alloying elements, a distinction between metallic and nonmetallic alloying elements must be made, since they have quite different diffusion rates and thus require varying sintering times to obtain a homogeneous structure.

In general, iron powder should contain very small amounts of carbon and other nonmetallic alloying elements, as these increase the hardness and decrease the compressibility (see Figure 6.1). The compressibility is measured as the density obtained for a compacting pressure of 400 MPa. The preferred method is to mix the powder with graphite (1 % graphite results in steel with 0.8 % C after sintering) so that good compressibility is retained.

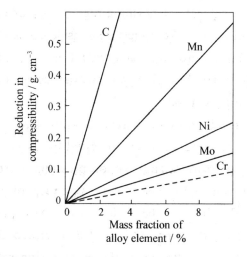

Figure 6.1 The influence of alloying elements on the compressibility of steel powders

Alloying by mixing the powder with the right amounts of metallic elements will require a very long sintering time to achieve a homogeneous structure. To reduce the sintering time required, prealloyed powders are preferred, as the reduction in compressibility is not very severe. If reduction-powders are used, alloying with metallic elements is normally based on partial alloying, which required a special alloying stage. The special pure powder is mixed with the alloying elements and heat-treated so that an incomplete diffusion into the base powder has taken place. During the final sintering, after compaction, the diffusion is completed. If atomization powders are used, a regular or true alloying is obtained, as the alloyed powder is produced directly from the liquid state. Such powders are called prealloyed powders.

In the past decade, a rapid increase in consumption of partial alloyed (reduction) powders and prealloyed (atomization) powders has taken place. By using these powders, excellent mechanical properties of the components can be obtained with tensile strengths in the range of 400~1 000 MPa, and even values of 1 500 MPa can be reached with special and more expensive powders. The alloying elements are primarily Cu, Ni, Mo and Mn. Stainless steel powders are being employed at an increasing rate. The development of powders leading to high-strength components considerably increases the potential market for powder metallurgy.

The discussion above focused on iron and steel powders, but it must be emphasized that a wide spectrum of nonferrous metal powders are available, including brass, bronze, aluminum, nickel and zinc powders.

Pressing or compacting of powders: the technology of pressing or compacting is a broad and complicated subject, requiring a high degree of engineering ingenuity. Therefore, the description given here must be considered as elementary.

Background: in this section the background for the effect of the different powders on die design is discussed briefly before the various die design principles (i. e., compacting methods) are discussed.

The component is specified by the desired density, strength, tolerances, and so on, and the powder is specified by its compressibility curve, that is, its density as function of compacting pressure (double-action pressing[2]) (see Figure 6.2a). The apparent or filling density is 2.4 g/cm^3, compaction ratio of 2 (resulting in half the original height), the density will be 4.8 g/cm^3. For practical purposes, the compaction ratio must be in the range of 2.5~2.8, corresponding to densities in the range of 6~7 g/cm^3. Figure 6.2b shows the compression ratio versus compacting pressure. From Figure 6.2a new curve showing the punch motion within the die as a function of the compacting pressure can be constructed based on the desired density and for 100 % punch motion. Such a curve shows that about 85 % of the motion is already carried out at compacting pressure of 200 MPa. For the remaining 15 % of the motion, the compaction pressure must increase from 200 to about 800~1 000 MPa, which means that the compaction press in required is only to provide high compacting pressure over a very short travel length.

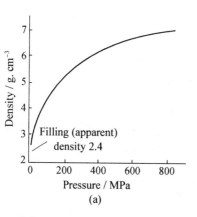 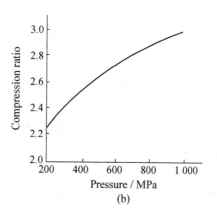

Figure 6.2　Density and compression ratio as a function of compacting pressure for iron powder

Approximately 90 % of the industrially applied powder components have densities in the range of 5.7~6.8 g/cm³, but in the past few years, the application of components having densities in the range of 7.0~7.2 g/cm³ has increased, these components have excellent mechanical properties. This has become economical due to the development of better die materials with higher wear resistance and powders with high compressibility.

The filling of the die cavity is generally achieved through volume dosing where the powder flows into the cavity and excess powder is scraped off, giving tolerances of ±1 %. If higher accuracy is required, dosing by weight must be used, but this is more tedious.

When the powder has been compacted, it must be ejected from the die. The ejection phase must be considered carefully, since fracture may arise at weak points or weak sections, when the elastic energy is released, or when force act over a small fraction of the powder surface. To obtain optimal production, the designer of the component must analyze both the compaction and ejection phases carefully before deciding on the final geometry.

Notes

① Hoeganaes method 海格纳斯法　　② double-action pressing 双向压制

Vocabulary

atomization [ˌætəʊmaɪˈzeɪʃən] *n.* 雾化，[分化]原子化

compaction [kəmˈpækʃən] *n.* 压紧；精简；密封；凝结

compressibility [kəmˌpresəˈbɪlətɪ] *n.* 压缩性；压缩系数；压缩率

ejection [ɪˈdʒekʃən] *n.* 喷出；排出物

electrolytic [ɪˌlektrəʊˈlɪtɪk] *adj.* 电解

Chapter 6 Powder Metallurgy

的;电解质的;由电解产生的
granular ['grænjʊlə] *adj.* 颗粒的;粒状的
lubrication [ˌluːbrɪ'keɪʃən] *n.* 润滑;润滑作用
metallurgy [mɪ'tælədʒɪ] *n.* 冶金;冶金学;冶金术
porosity [pɔː'rɒsɪtɪ] *n.* 有孔性,多孔性
prealloy [pˈrɪəlɒɪ] *n.* 预制合金

recommendation [ˌrekəmen'deɪʃ(ə)n] *n.* 推荐;建议;推荐信
stabilization [ˌsteɪbɪlaɪ'zeɪʃən] *n.* 稳定;稳定化
stearate ['stɪəreɪt] *n.* [有化]硬脂酸盐
stearic [sti'ærɪk] *adj.* 硬脂的;硬脂酸的;似硬脂的;十八酸的
tedious ['tiːdɪəs] *adj.* 沉闷的;冗长乏味的

Exercises

1. Translate the following Chinese phrases into English

(1) 粉末冶金   (2) 形状的稳定   (3) 钨粉
(4) 不规则颗粒   (5) 颗粒大小分布   (6) 不锈钢粉末
(7) 排出期   (8) 预合金粉末

2. Translate the following English phrases into Chinese

(1) granular materials   (2) reduction process
(3) refining process   (4) electrolytic deposition
(5) homogeneous structure   (6) high-strength component
(7) final geometry

3. Translate the following Chinese sentences into English

(1) 这种细粒度是通过在熔融操作过程中添加成核剂(通常是 TiO_2 和 ZrO_2)而形成的,紧接着是可控结晶。

(2) 氧化还原反应为什么必须同时进行?

(3) 第二部分讨论了纳米复合材料的制备工艺。采用两种制备方法:粉末冶金法和机械合金化法。

(4) 采用粉末烧结的方法制备了孔隙定向排列多孔高温合金材料。

(5) 在这篇论文中,介绍了温压工艺及其在硬质合金制备中的应用研究情况。

4. Translate the following English sentences into Chinese

(1) Metallurgy is a domain of materials science that studies the physical and chemical behavior of metallic elements, their intermetallic compounds, and their compounds, which are called alloys.

(2) When nanopowders were used in powder alloy process, it can decrease melting of the materials and the sintering temperature of which is below far-off that of normal.

(3) With the same atomized parameters, the porosity amount in large size powder is more than that in small size powder.

(4) SBP is high-tech product, which resolves the hard problem of spray granulation in hard metal industry.

(5) The tests in use show that this kind of iron powder has good compressibility, and is suitable for manufacturing parts with high density and complicated form.

(6) The prefix "nano" is from the Greek word "nano" and it means dwarf. Nanometer is a length unit. A nanometer (nm) equals a billionth of a meter (1 nm = 1×10^{-9} m).

5. Translate the following Chinese essay into English

粉末冶金也被称为粉末成形,它是一种利用金属粉末制作零部件的工艺。最初,这种工艺用来代替具有较高熔点的难熔金属的铸造。技术的发展使得很经济地生产一种产品成为可能。如今,粉末冶金在金属加工领域有着很重要的地位。

粉末冶金加工技术包括三步:混合金属或合金粉末,于室温条件下在模具里压实粉末,在一个环境可控的熔炉中烧结或高温加热模型使得金属微粒结合在一起。

6. Translate the following English essay into Chinese

Sintering involves heating of the green compact at high temperatures in a controlled atmosphere. Sintering increases the bond between the particles and therefore strengthens the powder metal compact. Sintering temperature is usually 0.6 to 0.8 times the melting point of the powder. In case of mixed powders of different melting temperature, the sintering temperature will usually be above the melting point of one of the minor constituent and other powders remain in solid state. The important factors governing sintering are temperature, time and atmosphere.

扫一扫,查看更多资料

Chapter 7 Casting Process and Forming Operation

7.1 Casting Process

7.1.1 Introduction

In casting, the liquid material is poured into a cavity (die or mold) corresponding to the desired geometry. The shape obtained in the liquid material is now stabilized, usually by solidification, and can be removed from the cavity as a solid component.

Casting is the oldest known process to produce metallic components. The main stages, which are not confined to metallic materials alone but are also applicable to some plastics, porcelain, and so on, are production of a suitable mold cavity; the melting of the material; pouring the liquid materials into the cavity; stabilization of the shape by solidification, chemical hardening, evaporation, and so on; removal or extraction of the solid component from the mold; and cleaning the component.

In principle, no limits exist regarding the size or geometry of the parts, which can be produced by casting. The limitations are set primarily by the material properties, the melting temperature, the properties of the mold material (mechanical, chemical and thermal) and the material's production characteristics (i. e., whether it is used only once or many times).

Normally, the term "casting" is applied to metals but in general, the principal stages and many of the characteristic problems are the same for most materials that can be shaped from the liquid state. The term "casting" should be treated more broadly, allowing the carryover of new ideas from one field to another (foundry industry, glass industry, plastic industry, etc.). Having a good knowledge of one field allows an easier understanding of another field.

Chapter 7 Casting Process and Forming Operation

The differences between the many casting processes are mainly due to the mechanical and thermal properties of the work and mold material, the acceptable working temperature of the mold, the cooling method and cooling rate of the workpiece, the radiation of heat from the work and the mold material, the chemical reaction between the molten metal and the mold, the solubility of gas in the work material, and functional requirements of the component.

In this context the discussion will be confined to the casting of metals, but it should be remembered that the principles are generally applicable to most materials that can be melted.

Casting processes are important and extensively used manufacturing methods, enabling the production of very complex or intricate parts in nearly all types of metals with high production rates, average to good tolerances and surface roughness, and good material properties. The competitiveness of the casing processes is based primarily on the fact that casting allows the elimination of substantial amounts of expensive machining often required in alternative production methods.

As mentioned, many different casting processes have been developed. The names associated with the processes may be related to the type of mold (nonpermanent, permanent) or to the mold material or the pouring method (gravity, high pressure, low pressure). Furthermore, the application of the names is not always consistent, which sometimes causes confusion. Table 7.1 shows the major casting processes classified according to the different characteristics. The most commonly used names are given, but if doubt about them arises, they can be identified by their characteristics. The individual processes are described later.

Table 7.1 Some characteristics of major casting processes

Type mold	Mold material	Pouring principle	Pattern material	Process name	Grouping
Nonpermantent (single-purpose)	Sand (green)	Gravity	Wood, metal, plastics	Green sand, dry sand core sand casting	Sand casting
Permanent	Alloy steels Graphite, Steel cast iron	High pressure	—	Die casting	Permanent (metallic) mold casting
		Low pressure	—	Low pressure (permanent mold) casting	
		Gravity	—	Nonpressure-gravit permanent mold casting	

Chapter 7 Casting Process and Forming Operation

(to be continued)

Type mold	Mold material	Pouring principle	Pattern material	Process name	Grouping
Nonpermantent (single-purpose)	Nonmetallic(sand, plaster, ceramics, etc.)	Gravity (low pressure)	Metal, wax, plastic, rubber Wax	Shell mold cast Plaster mold casting Ceramic shell mold casting "Low wax" casting (investment casting)	Precision casting
Nonpermantent permanent	Nonmetallic Metallic	Centrifugal forces	—	Centrifugal casting	Centrifugal casting

7.1.2 Sand Casting

Sand casting is characterized by economical and operationally simply. For sand casting, ordinary sand is always used as the mold material. The two-piece mold[①] is usually formed by packing sand around a model which has the desired shape. In addition, a gating system is usually incorporated into the mold in order to drive the molten metal into the cavity more fluently. So the internal casting defects could be minimized. Though convenient, sand casting has several intrinsic drawbacks, Such as comparatively high defective rate and high surface roughness.

7.1.3 Investment Casting

In investment casting (also called lost "wax casting" or "precision casting") a pattern of wax is used.

The main stages in the investment casting process are:

(1) Production of a master pattern (normally in metal but wood or plastic is sometimes used), used to produce a master die (low-melting-point alloys or steel).

(2) Production of wax patterns by pouring or injection of wax into the master die.

(3) Assembly of wax patterns and a common gating system with a sprue (a soldering iron can be used), called a cluster if several patterns are to be united.

(4) Coating of the pattern assembly with a thin layer of investment material (dipping in thin slurry of fine-grained silica).

(5) Production of the final investment by placing the coated pattern assembly into a flask and pouring investment material around (vibrated to remove entrapped air, etc.).

(6) Drying and hardening for several hours.

(7) Melting the wax pattern assembly by warming the mold and inverting it to allow the wax to flow out.

(8) Heating the mold to higher temperatures (850~1 000 ℃) to drive off moisture and volatile matter.

(9) Preheating the mold to 500~1 000 ℃ (facilitating flow of the molten metal to thin sections to give better dimensional control).

(10) Pouring the metal (by gravity, pressure or evacuation of the mold).

(11) Removal of the casting from the mold after solidification.

Polystyrene or frozen mercury can also be used as the pattern material; rubber may be used for permanent patterns whenever it is possible to extract it after investment.

Fine silica (bonded by tetraethyl silicate or sodium ammonium phosphate), plaster (for low pouring temperatures, for example, with magnesium, aluminum and some copper alloys) or ceramics can be used as mold material. The use of plaster molds and ceramic molds has increased rapidly during recent years. The molds may be reinforced by fibers. In many cases, shell molds are used (produced by dipping about five times), which reduces the quantity of investment material, eliminates flasks, reduces firing time, and simplifies removal of the casting from the mold.

The investment process can be used to produce castings in all ferrous and nonferrous alloys and is important in the casting of special metals such as unmachinable alloys and radioactive metals.

The main advantages of investment casting are: the production of very complicated shapes even in high melting temperature alloys (this includes thin sections (about 0.4 mm), undercuts, *etc*.), very fine details, exceptionally good surface finish and very high dimensional accuracy (0.003~0.005 per dimensional unit, cm).

The labor costs in investment casting are high. The pattern costs are also high; consequently, the process is used mainly to produce components that require the special characteristics of the process (good surfaces, tolerances, high complexity, *etc*.). Examples are metals that are difficult to machine or to deform plastically.

7.1.4 Die Casting

Die casting is characterized by a permanent metal mold and high pouring or injection pressure. The injection pressure, under which solidification also takes

Chapter 7 Casting Process and Forming Operation

place, may vary from 2 to 300 MPa; the usual range is 10~50 MPa. Two different die casting methods are employed: the hot-chamber method[②] and the cold-chamber method[③]. The principal distinguishing feature is the location of the melting pot, which also reflects the final design of the equipment. In the hot-chamber method, the melting pot is included in the machine and the injection cylinder is immersed in the molten metal (see Figure 7.1a and Figure 7.1b). Figure 7.1a shows the metal being forced by air into the die (pressure of 0.5~5 MPa) and Figure 7.1b shows the metal being forced into the die by a plunger (activated by air or hydraulic pressure), resulting in injection and solidification pressure in the range of 10~40 MPa. Hot-chamber die casting is used mainly for the casting of alloys of zinc, tin, lead and magnesium.

The cold-chamber process has a separate melting furnace, and the molten metal is transferred from the furnace to the cold-chamber machine by hand or mechanically (see Figure 7.1c). In the machine, the metal is forced into the die by a hydraulically activated plunger. The injection and solidification pressure is in the range of 30~150 MPa. Machines of 25 MN plunger force and more are available, allowing the casting of components of up to 50 kg. The cold-chamber method is used mainly for brass, bronze, aluminum and magnesium castings.

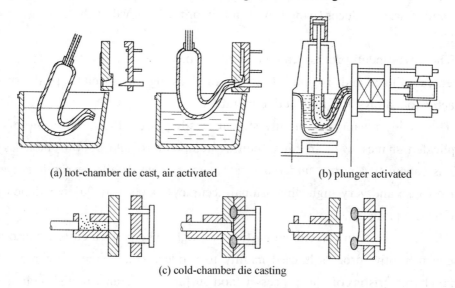

(a) hot-chamber die cast, air activated   (b) plunger activated

(c) cold-chamber die casting

Figure 7.1 Die casting

The die casting process is rapid (production rates of up to 1 000 casting per hour) and it gives smooth surfaces, good dimensional accuracy (0.002 to 0.003 per dimensional unit, cm/cm of length) and thin sections (0.6 mm in zinc, 0.09 mm in aluminum, and 1.25 mm in magnesium, brass and bronze). The draft necessary varies between 0.125 mm and 0.35 per dimensional unit (cm/cm), depending on the material.

Die casting requires, in general, no machining except for the drilling of holes and threading. Flash and fins must be removed.

The dies for die casting are made from heat-resistant steel and are water-cooled through internal channels. For large castings, a single-cavity die④ is used and for small castings, multiple cavity dies are used, often built up from inserts.

7.2 Forming Operations

7.2.1 Introduction

Forming operations are the techniques in which the shape of a metal piece is changed by plastic deformation, including forging, rolling, extrusion, drawing, and so on.

7.2.2 Forging

Forging is a process that deforming a single piece of hot metal by means of metal forming machinery. Forgings are classified as cold and hot forging, according to the ingot temperature during process. For cold forging, the ingot is usually maintained at room temperature. For hot forging, on the other hand, the temperature is usually higher than the recrystallization temperature of the ingot. Hot forging is more common and includes closed-die and open-die forging. For closed-die, a force is brought to bear on a metal slug or perform placed between two or more die halves. The metal flows plastically into the cavity formed by the die and hence changes in shape to its finished shape in Figure 7.2a. Open-die forging is performed between flat dies with no pre-cut profiles. The dies do not confine the metal laterally during forging. Deformation is achieved through movement of the workpiece relative to the dies. Parts up to thirty metres in length can be hammered or pressed into shape in this way. Open-die forging comprises many process variations, enabling: an extremely broad range of shapes and sizes to be produced in Figure 7.2b. The forged articles have excellent mechanical properties, combining fine grain structure with strengthening through strain hardening. For example, the porosity of the as-cast articles can be removed by forging.

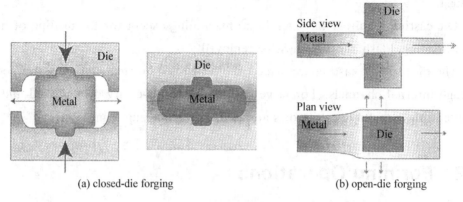

(a) closed-die forging   (b) open-die forging

Figure 7.2 Schematic of hot forging

7.2.3 Rolling

Rolling is one of the most widely used deformation process. It consists of passing metal between two rollers, which exert compressive stresses, reducing the metal thickness. Where simple shapes are to be made in large quantity, rolling is the most economical process. Rolled products include sheets, structural shapes and rails as well as intermediate shapes for wire drawing or forging. Circular shapes, "I" beams and railway tracks are manufactured using grooved rolls in Figure 7.3.

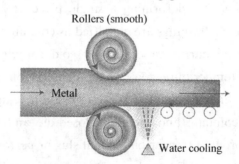

Figure 7.3 Schematic of rolling

Initial breakdown of an ingot or a continuously cast slab is achieved by hot rolling. Mechanical strength is improved and porosity is reduced. The worked metal tends to oxidize leading to scaling which results in a poor surface finish and loss of precise dimensions. A hot rolled product is often pickled to remove scale, and further rolled cold to ensure a good surface finish and optimize the mechanical properties for a given application. Cold rolling is often used in the final stages of production. Sheets, strips and foils are cold rolled to attain dimensional accuracy and high quality surface finishes.

7.2.4 Extrusion

In extrusion, a bar of metal is forced from an enclosed cavity via a die orifice by a compressive force applied by a ram. Since there are no tensile forces, high deformations are possible without the risk of fracture of the extruded material. The extruded article has the desired, reduced cross-sectional area, and also has a good surface finish so that further machining is not needed. Extrusion products include rods and tubes with varying degrees of complexity in cross-section in Figure 7.4.

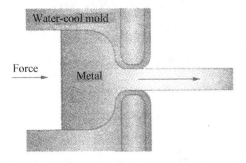

Figure 7.4 Schematic of extrusion

Hot extrusion is carried out at a temperature of approximately 0.6 T_m and the pressures required the range from 35 to 700 MPa. While, cold extrusion is performed at temperatures significantly below the melting temperature of the alloy being deformed, and generally at room temperature.

7.2.5 Drawing

Drawing is the pulling of a metal piece through a die by means of a tensile force applied to the exit side. A reduction in cross-sectional area results, with a corresponding increase in length. A complete drawing apparatus may include up to twelve dies in a series sequence, each with a hole a little smaller than the preceding one. In multiple-die machines, each stage results in an increase in length and therefore a corresponding increase in speed is required between each stage.

Metals can be formed to much closer dimensions by drawing than by rolling. Shapes ranging in size from the finest wire to those with cross-sectional areas of many square centimetres can be commonly drawn. Drawn products include wires, rods and tubing products. Large quantities of steel and brass are cold drawn. Seamless tubing can be produced by cold drawing when thin walls and very accurate finishes are required in Figure 7.5.

Chapter 7 Casting Process and Forming Operation

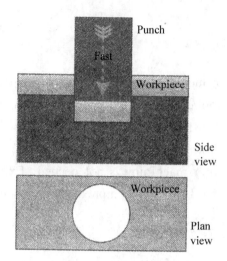

Figure 7.5 Schematic of drawing

7.2.6 Stamping

Stamping is used to make high volume parts such as aviation or car panels or electronic components. Mechanical orhydraulic powered presses stamp out parts from continuous sheets of metal or individual blanks. The upper die is attached to the ram and the lower die is fixed. Whereas mechanical machinery transfers all energy as a rapid punch, hydraulic machinery delivers a constant, controlled force.

Notes

① two-piece mold 两半模具
② hot-chamber method 热室法
③ cold-chamber method 冷室法
④ single-cavity die 单穴模

Vocabulary

ammonium [əˈməʊnɪəm] n. [无化]铵；氨盐基
aviation [ˌeɪvɪˈeɪʃ(ə)n] n. 航空；飞行术；飞机制造业
cavity [ˈkævɪtɪ] n. 腔；洞，凹处
centimeter [ˈsentɪmiːtə(r)] n. [计量]厘米；[计量]公分
drawing [ˈdrɔː(r)ɪŋ] n. 图画；牵引；素描 v. 绘画；吸引（draw 的 ing 形式）；拖曳
evacuation [ɪˌvækjʊˈeɪʃ(ə)n] n. 疏散；撤离；排泄
extrusion [ɪkˈstruːʒn] n. 挤出；推出；赶出；喷出
foundry [ˈfaʊndrɪ] n. 铸造，铸造类；[机]铸造厂
hydraulic [haɪˈdrɔːlɪk] adj. 液压的；水力的；水力学的

Chapter 7 Casting Process and Forming Operation

ingot [ˈɪŋɡət] n. 锭；铸块
moisture [ˈmɔɪstʃə] n. 水分；湿度；潮湿；降雨量
phosphate [ˈfɒsfeɪt] n. 磷酸盐；皮膜化成
plunger [ˈplʌn(d)ʒə] n. [机]活塞；潜水者；跳水者；莽撞的人
polystyrene [ˌpɒlɪˈstaɪriːn] n. [高分子]聚苯乙烯
porcelain [ˈpɔːs(ə)lɪn] n. 瓷；瓷器 adj. 瓷制的；精美的
ram [ræm] n. 撞锤；撞击装置；[机]活塞 v. 撞击；填塞
rolling [ˈrəʊlɪŋ] n. 旋转；动摇；轰响 adj. 旋转的；起伏的；波动的
seamless [ˈsiːmlɪs] adj. 无缝的；无缝合线的；无伤痕的
stamping [ˈstæmpɪŋ] n. 冲击制品 v. 冲压
tetraethyl [ˌtetrəˈiːθaɪl] n. 四乙基的
volatile [ˈvɒlətaɪl] n. 挥发物；有翅的动物 adj. [化学]挥发性的；不稳定的；爆炸性的；反复无常的

Exercises

1. Translate the following Chinese phrases into English

(1) 铸造法   (2) 砂型铸造   (3) 细硅粉
(4) 压铸   (5) 耐热钢

2. Translate the following English phrases into Chinese

(1) forming operation   (2) intrinsic drawback
(3) high surface roughness   (4) injection pressure
(5) ingot temperature

3. Translate the following Chinese sentences into English

(1) 适当地选用模具材料和模具制造技术，在很大程度上决定了成型模具的使用寿命。
(2) 因为有高的金属注射压力，压力铸造具有将各种气孔压缩得比其它任何铸造方法都小的能力。
(3) 制造过程很简单：把融熔的塑料注入一个模具然后它就成了。
(4) 这样制造商就能使用现有的模具、部件、装配线和培训等等资源。
(5) 注射压力增加，生坯密度和抗弯强度都增大。

4. Translate the following English sentences into Chinese

(1) For cost effective reasons protoypes, with dimensions close to the final product, and reforming procedures are used, which can minimize the mechanical finishing of the construction unit.

(2) It can thus be seen that the most generous venting of the mold will not ensure completely sound casting if the metal does not cross the cavity as a solid front.

(3) Additionally, it was found effective to match the gate thickness at any one

point to distance that the metal must flow to reach the overflow pocket.

5. Translate the following Chinese essay into English

砂型铸造的第一步是制作铸造模具。在一次性模具工艺中,每一次铸造都要进行这一步。将沙子塞进每一半的模具中就制成了砂模。沙子被堆放在木模周围,这个木模的外部轮廓和铸件的外部轮廓相同。当木模被取下的时候,用于形成铸件的空洞保存下来。任何不能用木模组成的铸件内部特征都用由沙子做成的型芯构成,这些型芯是在模具形成之前做好的。模具制造时间包括木模定位、灌注沙子和取出木模。模腔的表面需要润滑,以利于铸件的移除。润滑剂的使用同样能改善金属液的流动,提高铸件的表面光洁度。

6. Translate the following English essay into Chinese

Impression die forging, also called closed die forging is the deformation of metal at forging temperature within one or more die impressions or cavities. It is performed both in presses and hammers. Work pieces may be round or rectangular in cross section or flat disks, and the dies are sometimes interrelated heated to minimize chilling and cooling in the work piece. For simple shapes, impression die forging can be performed in a single prestrike. More often, however, several strokes of different forces used with dies having several impressions for sequential forming and final operations.

扫一扫,查看更多资料

Chapter 8 Welding

8.1 Introduction

Welding differed from rivet and screws fasten, is a joint method in which atomic bonding occurs between two metal pieces. Welding is mainly employed to manufacture metallic hardware. For example, boiler, pressure container, pipeline, ship craft, vehicle, aircraft, *etc.* are always produced by welding. Welding is needed in almost all industrial fields. Over 60 % of annually steel output is produced by welding in some important industrialized nations.

Welding has been extensively used due to a series of advantages. Firstly, welding can produce articles with excellent joint properties. The welding seam is characterized by good leak tightness, conductivity, wear and corrosion-resistance[①]. Secondly, compared with rivet, welding is more economical. About 10~20 % metallic materials can be saved if rivet is taken place by welding. Thirdly, the weight of metallic hardware can be reduced by welding, which is quite important for launch vehicles like ship craft, rocket, *etc*. Finally, the manufacturing procedure can be significantly simplified by welding, especially for heavy and complicated pieces.

8.2 Definition and Classification of Welding Processes

Welding is essential for the manufacture of a range of engineering components, which may vary from very large structures such as ships and bridges to miniature components for microelectronic applications.

Several alternative definitions used to describe weld, for example: A union between two pieces of metal rendered plastic or liquid by heat or pressure or both. A filler metal with a melting temperature of the same order of that of the parent

Chapter 8 Welding

metal may or may not be used.

Or alternatively: A localized coalescence of metals or nonmetals produced either by heating the materials to the welding temperature with or without the application of pressure, or by the application of pressure alone, with or without the use of a filler metal.

Based on these definitions welding processes may be classifiedinto those which rely on the application of pressure and those which used elevated temperatures to achieve the bond. The most important processes are shown in Figure 8.1.

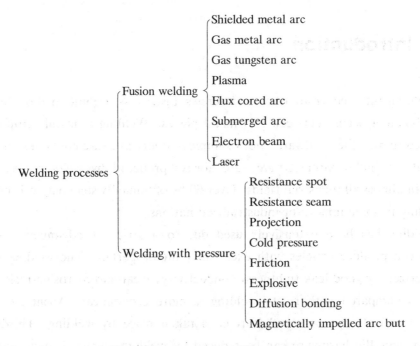

Figure 8.1 Important welding processes

8.3 Fusion Welding

The most widely used welding processes rely on fusion of the components at the joint line. In fusion welding, a heat source melts the metal to form a bridge between the components.

A widely used heat source is electric arc, as shown in Figure 8.2.

The molten metal must be protected from the atmosphere absorption of oxygen and

Figure 8.2 Welding arc

nitrogen leads to a poor quality weld. Air in the weld area can be replaced by a gas which does not contaminate the metal, or the weld can be covered with a flux.

A large number of fusion welding processes and techniques are available. No process is universally best. Each has its own special attributes and must be matched to application. Choosing the most suitable process requires consideration of a number of factors, including type of metal, type of joint, material thickness, production constraints, equipment availability, labour availability, labour costs, costs of consumables, health, safety and the environment consideration.

8.3.1 Shielded Metal Arc Welding

Shielded metal arc welding (SMAW) process, shown in Figure 8.3, is also known as manual metal arc welding (MMAW) in Europe.

Figure 8.3 Shielded metal arc welding

It has for many years been one of the most common techniques applied to the fabrication of steels. The process uses an arc as the heat source and shielding is provided by gases generated by the decomposition of the electrode coating material and by the slag produced by the melting of mineral constituents of the coating. In addition to heating and melting the parent material the arc also melts the core of the electrode and thereby provides filler material for the joint. The electrode coating may also be used as a source of alloying elements and additional filler material. The flux and electrode chemistry may be formulated to deposit wear- and corrosion-resistant layers for surface protection.

Significant features of the process are: equipment requirements are simple; a large range of consumables are available; the process is extremely portable; the operating efficiency is low; it is labour intensive.

For these reasons the process has been traditionally used in structural steel fabrication, shipbuilding and heavy engineering as well as for small batch production and maintenance.

8.3.2 Gas Metal Arc Welding

Gas metal arc welding (GMAW), shown in Figure 8.4 and Figure 8.5, is also known as metal inert gas (MIG) or metal active gas (MAG) welding in Europe.

Figure 8.4 Gas metal arc welding

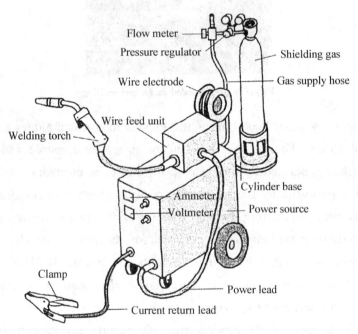

Figure 8.5 Gas metal arc welding system

Gas metal arc welding uses the heat generated by an electric arc to fuse the joint area, the arc is formed between the tip of a consumable, continuously fed filler wire and the work piece and the entire arc area is shielded by an inert gas. The principle of operation is illustrated in Figure 8.6.

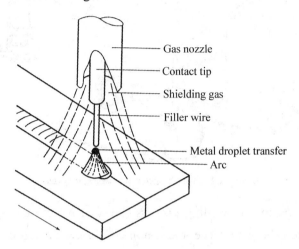

Figure 8.6　The operating principle of gas metal arc welding

Some of the more important features of the process are summarized below: low heat input (compared with SMAW and SAW); continuous operation; high deposition rate; no heavy slag—reduced post-weld cleaning; low hydrogen—reduces risk of cold cracking.

Depending on operation mode of the process it may be used at low currents for thin sheet or positional welding.

The process is used for joining plain carbon steel sheet from 0.5 to 2 mm thick in the following applications: automobile bodies, exhaust systems, storage tanks, tubular steel furniture, heating and ventilating ducts. The process is applied to positional welding of thick plain carbon and low alloy steels in the following areas: oil pipelines, marine structures and earth-moving equipment[②]. At higher currents high deposition rates may be obtained and the process is used for downhand and horizontal-vertical welds in a wide range of materials include earth-moving equipment, structural steelwork, weld surfacing with nickel or chromium alloys, aluminum alloy cryogenic vessels and military vehicles.

8.3.3　Gas Tungsten Arc Welding

Gas tungsten arc welding (GTAW), shown in Figure 8.7, is also known as tungsten inert gas (TIG) welding in most of Europe.

Chapter 8 Welding

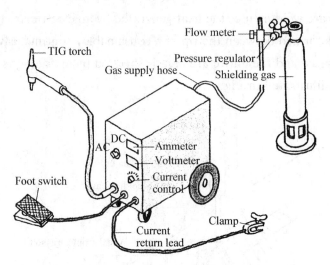

Figure 8.7 Gas tungsten arc welding system

In the gas tungsten arc welding process the heat generated by an arc which is maintained between the workpiece and a non-consumable tungsten electrode is used to fuse the joint area. The arc is sustained in an inert gas which serves to protect the weld pool and the electrode from atmospheric contamination. The principle of operation is illustrated in Figure 8.8.

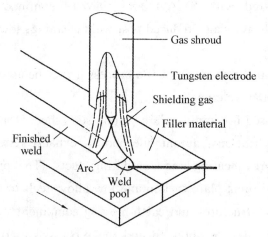

Figure 8.8 The operating principle of gas tungsten arc welding

The process has the following features: it is conducted in a chemically inert atmosphere; the arc energy density is relatively high; the process is very controllable; joint quality is usually high; deposition rates and joint completion rates are low.

The process may be applied to the joining of a wide range of engineering materials including stainless steel, aluminum alloys and reactive metals such as titanium. These features of the process lead to its widespread application in the aerospace, nuclear reprocessing and power generation industries as well as in the fabrication of chemical process plant, food processing and brewing equipment.

8.3.4 Plasma Welding

The arc used in TIG welding can be converted to a high energy jet by forcing it through a small hole in a nozzle. This constricts the arc and forms the plasma jet. Plasma welding uses the heat generated by the constricted arc to fuse the joint area, and the arc is formed between the tip of a non-consumable electrode[③] and either the workpiece or the constricting nozzle, as shown in Figure 8.9. A wide range of shielding gases is used depending on the mode of operation and the application.

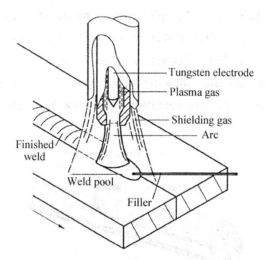

Figure 8.9 Plasma welding

In the normal transferred arc mode the arc is maintained between the electrode and workpiece; the electrode is usually the cathode and the workpiece is connected to positive side of the power supply. In this mode a high energy density is achieved and the process may be used effectively for welding.

Plasma welding relies on a special technique known as keyhole. First a hole is pierced through the joint by plasma arc. As the torch is moved along joint, metal melts at front of the hole, swirls to the back and solidifies.

The features of the process depend on the operating mode and the current, but in summary the plasma process has the following characteristics: good low-current arc stability; improved directionality compared with TIG; improved melting efficiency compared with TIG; possibility of keyhole welding.

These features of the process make it suitable for a range of applications including the joining of very thin materials, the encapsulation of electronic components and sensors, and high-speed longitudinal welds on strip and pipe.

8.3.5 Flux-Cored Wire Welding

Flux-cored wires[④] consist of a metal outer sheath filled with a combination of mineral flux and metal powders, as shown in Figure 8.10. The flux-cored wires welding (FCWW) process is operated in a similar manner to GMAW welding and the principle is illustrated in Figure 8.11. The most common production technique used to produce the wire involves folding a thin metal strip into a U shape, filling it with the flux constituents, closing the U to form a circular section and reducing the diameter of the tube by drawing or rolling. Alternative configurations, shown in Figure 8.12, may be produced by lapping or folding the strip or the consumable may be made by filling a tube flux followed by a drawing operation to reduce the diameter. Typical finished wire diameters range from 3.2 to 0.8 mm.

Flux-cored wires offer the following advantages: high deposition rates; alloying addition from the flux core; slag shielding and support; improved arc stabilization and shielding.

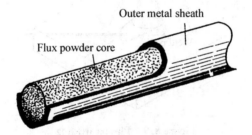

Figure 8.10 Construction of a flux cored wire

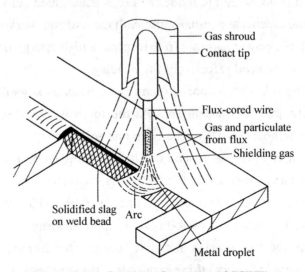

Figure 8.11 Principle of operation of FCWW

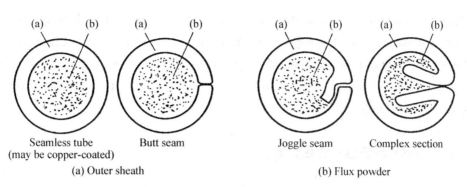

Figure 8.12　Alternative configurations of flux-cored wires

8.3.6　Submerged Arc Welding

Submerged arc welding (SAW), shown in Figure 8.13, is a consumable electrode arc welding process in which the arc is shielded by a molten slag and the arc atmosphere is generated by decomposition of certain slag constituents. The filler material a continuously fed wire and very high melting and deposition rate are achieved by using high current (*e. g.*, 1 000 A) with relatively small-diameter wires (*e. g.*, 4 mm).

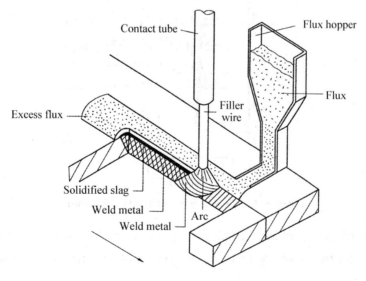

Figure 8.13　Submerged arc welding

The significant features of the process are: high deposition rates; automatic operation; no visible arc radiation; no visible range of flux/wire combinations; difficult to use positionally; best for thicknesses above 6 mm.

The main applications of submerged arc welding are on thick-section plain carbon and low-alloy steels and it has been used on power generation plant, nuclear containment, heavy structural steelwork, offshore structures and shipbuilding. The process is also used for high-speed welding of simple geometric seams in thinner sections, for example in fabrication of pressure containers for liquefied petroleum gas. Like shielded metal arc welding, with suitable wire/flux combinations the process may also be used for surfacing.

8.3.7　Electron Beam Welding

A beam of electrons may be accelerated by a high voltage to provide a high-energy heat source for welding, as shown in Figure 8.14. The power density of electron beams is high (10^{10} to 10^{12} W/m^2) and keyhole welding is the normal operating mode. The problem of power dissipation when the electrons collide with atmospheric gas molecules is usually overcome by carrying out the welding operation in a vacuum.

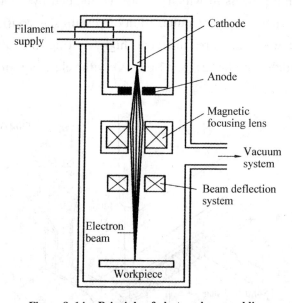

Figure 8.14　Principle of electron beam welding

Features of the process include: very high energy density; confined heat source; high depth to width ratio of welds; normally requires a vacuum; high equipment cost.

Applications of electron beam welding have traditionally included welding of aerospace engine components and instrumentation, but it may be used on a wide range of materials when high-precision and very deep penetration welds are required.

8.3.8 Laser Welding

The laser may be used as an alternative heat source for fusion welding. The focused power density of the laser can reach 10^{10} to 10^{12} W/m² and welding is often carried out using the "keyhole" technique.

Significant features of laser welding are: very confined heat source at low power; deep penetration at high power; reduced distortion and thermal damage; out-of-vacuum technique; high equipment cost.

These features have led to the application of lasers for microjoining of electronic components, but the process is also being applied to the fabrication of automotive and precision machine tool parts in heavy section steel.

8.4 Welding with Pressure

8.4.1 Resistance Spot Welding

Spot welding, as shown in Figure 8.15, is one of a group of resistance welding processes that involve the joining of two or more metal parts together in a localized area by the application of heat and pressure. The heat is generated within the material being joined by the resistance to the passage of a high current through the metal parts. Resistance heating at the contact surfaces causes local melting and fusion. High currents (typically 10 000 A) are applied for short durations and pressure is applied to the electrodes prior to the application of current and for a short time after the current has ceased to flow.

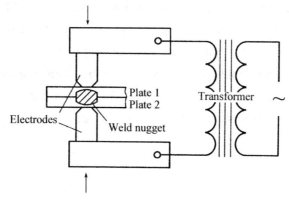

Figure 8.15 Resistance welding system

The process is used for joining sheet materials and uses shaped copper alloy electrodes to apply pressure and convey the electrical, current workpieces. Heat is developed mainly at the interface between two sheets, eventually causing the material being welded to melt, forming a molten pool, the weld nugget. The molten pool is contained by the pressure applied by the electrode tip and the surrounding solid metal.

Accurate control of current amplitude, pressure and weld cycle time are required to ensure that consistent weld quality is achieved but some variation may occur due to changes in the contact resistance of the material, electrode wear, magnetic losses or shunting of the current through previously formed spots. These "unpredictable" variations in process performance have led to the practice of increasing the number of welds from the design requirement to give some measure of protection against poor individual weld quality. To improve this situation significant developments have been, made in resistance monitoring and control, these allow more efficient use of process.

Features of the basic resistance welding process include: the process requires relatively simple equipment; it is easily and normally automated; once welding parameters are established it should be possible to produce repeatable welds for relatively long production runs.

The major applications of the process have been in the joining of sheet steel in the automotive and white-goods manufacturing industries.

8.4.2 Resistance Seam Welding

The seam welding process is an adaptation of resistance spot welding and involves making a series of overlapping spot welds by means of rotating copper alloy wheel electrodes to form a continuous leak tight joint. The electrodes are not opened between spots. The electrode wheels apply a constant force to the workpieces and rotate at a controlled speed. The welding current is normally pulsed to give a series of discrete spots, but may be continuous for certain high speed applications where gaps could otherwise occur between individual spots. Seam welding equipment is normally fixed and the components being welded are manipulated between the wheels. The process may be automated; it is illustrated in Figure 8.16.

Figure 8.16　Resistance seam welding

8.4.3　Resistance Projection Welding

Projection welding is a development of resistance spot welding. In spot welding, the size and position of the welds are determined by the size of the electrode tip and the contact point on the workpieces, whereas in projection welding the size and position of the weld or welds are determined by the design of the component to be welded. The force and current are concentrated in a small contact area which occurs naturally, as in cross wire welding or is deliberately introduced by machining or forming. An embossed dimple is used for sheet joining and a "V" projection or angle can be machined in a solid component to achieve an initial line contact with the component to which it is to be welded, see Figure 8.17.

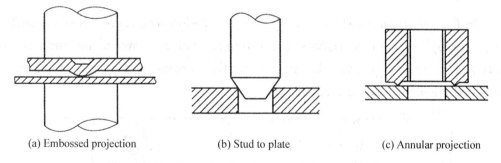

(a) Embossed projection　　　(b) Stud to plate　　　(c) Annular projection

Figure 8.17　Example of projection welding configurations

In sheet joining using embossed projection welds, a melted weld zone is produced, as in spot welding. However, when a solid formed or machined projection is used, a solid phase forge weld is produced without melting. The plastic deformation of the heated parts in contact produces a strong bond across the weld interface.

The process is well established and is applicable mainly to low carbon or microalloyed steels. The process is widely used on sheet metal assemblies in automotive and white goods industries for both sheet joining and attaching nuts and studs.

8.4.4 Cold Pressure Welding

If sufficient pressure is applied to the cleaned mating surfaces to cause substantial plastic deformation the surface layers of the material are disrupted, metallic bonds form across the interface and a cold pressure weld is formed.

The main characteristics of cold pressure welding are: the simplicity and low cost of the equipment; the avoidance of thermal damage to the material; most suitable for low-strength (soft) materials.

The pressure and deformation may be applied by rolling, indentation, butt welding, drawing or shear welding techniques. In general the more ductile materials are more easily welded.

This process has been used for electrical connections between small-diameter copper and aluminum conductors using butt and indentation techniques. Roll bonding is used to produce bimetallic sheets such as Cu/Al for cooking utensils, Al/Zn for printing plates and precious-metal contact springs for electrical applications.

8.4.5 Friction Welding

In friction welding, shown in Figure 8.18, a high temperature is developed at the joint by the relative motion of the contact surfaces. When the surfaces are softened a forging pressure is applied relative motion is stopped. Material is extruded from the joint to form an upset.

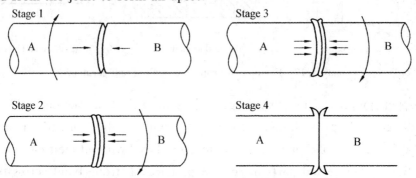

Figure 8.18 Friction welding

Chapter 8 Welding

Stage 1: A fixed, B rotated and moved into contact with A.

Stage 2: A fixed, B rotated under pressure, interface heating.

Stage 3: A fixed, forge pressure applied.

Stage 4: Relative motion stopped, weld formed.

The process may be divided into several operating modes in terms of the means of supplying the energy: continuous drive: in which the relative motion is generated by direct coupling to the energy source. The drive maintains a constant speed during the heating phase; stored energy: in which the relative motion is supplied by a flywheel which is disconnected from the drive during the heating phase.

The process may also be classified according to the type of motion as shown in Figure 8.19. Rotational motion is the most commonly used, mainly for round components where angular alignment of the two parts is not critical. If it is required to achieve a fix relationship between the mating parts angular oscillation may be used and for non-circular components the linear and orbital techniques may be employed.

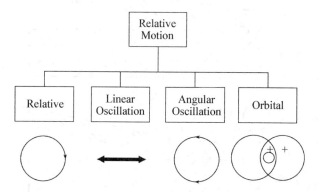

Figure 8.19 Friction welding: classification by motion type

Features of the process include: one-shot process for butt welding sections; suitable for dissimilar metals; short cycle time; most suited to circular sections; robust and costly equipment may be required.

The process is commonly applied to circular sections, particularly in steel, but it may also be applied to dissimilar metal joints such as aluminum to steel or even ceramic materials to metals. Early applications of the process included the welding of automotive stub axles but the process has also been applied to fabrication of high-quality aero-engine parts, duplex stainless steel pipe for offshore applications and nuclear components.

8.4.6 Explosive Welding

In explosive welding the force required to deform the interface is generated by an explosive charge. In the most common application of the process two flat plates are joined to form a bimetallic structure, shown in Figure 8.20. An explosive charge is used to force upper of "flyer" plate on to the base plate in such a way that a wave of plastic material at the interface is extruded forward as the plates join. For large workpieces considerable force is involved and care is required to ensure the safe operation of the process.

Features of the process include: one-shot process-short welding time; suitable for joining large surface areas; suitable for dissimilar thickness and metals joining; careful preparation required for large workpieces.

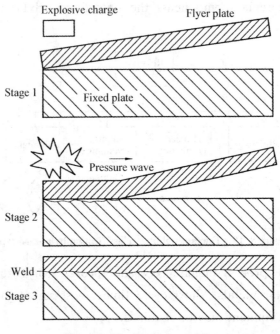

Figure 8.20 Explosive welding

8.4.7 Diffusion Bonding

In diffusion bonding the mating surfaces are cleaned and heated in an inert atmosphere. Pressure is applied to the joint and local plastic deformation is followed by diffusion during which the surface voids are gradually removed. The components to be joined need to be enclosed in a controlled atmosphere and the process of diffusion is time and temperature dependent. In some cases an intermediate material is placed between the abutting surfaces to form an interlayer.

Significant features of the process are: suitable for joining a wide range of materials; one-shot process; complex sections may be joined; vacuum or controlled atmosphere required; prolonged cycle tine.

8.4.8 Magnetically Impelled Arc Butt Welding

Magnetically impelled arc butt (MIAB) welding is a one shot, forge welding process which is predominantly used in the European automotive industry for rapidly joining circular and non-circular thin wall[5] (<5 mm) steel tubes. This machine tool based process is attractive to the mass production industries because of the short cycle times and reproducible quality.

The first stage of a MIAB weld is to force the two tubulars together whilst applying a DC welding current. They are then moved apart to a distance of 1~3 mm in order to strike an arc. This arc is rotated at high speed around the circumference of the weld interface using a static radial magnetic field which can be generated using permanent magnets or electromagnets. Arc rotation is sustained for a few seconds until the joint faces are heated to a high temperature or are molten, as shown in Figure 8.21.

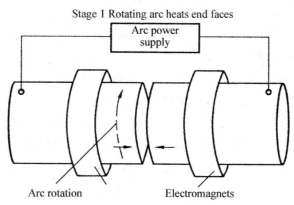

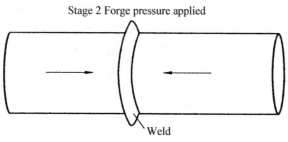

Figure 8.21　MIAB welding

Chapter 8　Welding

In the second stage of the process the tubulars are brought rapidly together under a pre-determined forging pressure and the arc is extinguished. The molten metal at the weld interface is expelled and a solid phase weld results from sustained forging pressure, which consolidates the joint. Typically weld cycle times range from 1~6 seconds depending on tube diameter.

Notes

① corrosion-resistance 耐蚀性
② earth-moving equipment 运土设备
③ non-consumable electrode 非自耗电极
④ flux-cored wires 药芯焊丝
⑤ non-circular thin wall 非圆形薄壁

Vocabulary

brewing ['bruːɪŋ] n. 酿造；酝酿；计划 v. 酿造；策划阴谋（brew 的 ing 形式）
bimetallic [baɪmɪ'tælɪk] adj. [材]双金属的；复本位制的
contaminate [kən'tæmɪneɪt] vt. 污染，弄脏
cryogenic [ˌkraɪə'dʒenɪk] adj. 冷冻的（副词 cryogenically）；低温学的；低温实验法的
dissimilar [dɪ'sɪmɪlə] adj. 不同的
downhand [hænddaʊn] n. 俯焊；平焊
duration [djʊ'reɪʃ(ə)n] n. 持续，持续的时间，期间
electrode [ɪ'lektrəʊd] n. 电极；电焊条
encapsulation [ɪnˌkæpsə'leɪʃən] n. 封装；包装
flywheel ['flaɪwiːl] n. [机]飞轮，惯性轮；调速轮
interlayer ['ɪntəleɪə] n. 夹层，隔层
keyhole ['kiːhəʊl] n. 钥匙孔，锁眼，钥孔形弹孔

miniature ['mɪnɪtʃə] n. 缩图；微型画；微型图画绘画术　adj. 微型的，小规模的　vt. 是……的缩影
nozzle ['nɒz(ə)l] n. 喷嘴；排气口；嘴
nugget ['nʌgɪt] n. 点焊熔核
petroleum [pə'trəʊlɪəm] n. 石油
pipeline ['paɪplaɪn] n. 管道；输油管；传递途径
plasma ['plæzmə] n. [等离子]等离子体
rivet ['rɪvɪt] n. 铆钉　vt. 铆接；固定
robust [rəʊ'bʌst] adj. 强健的；稳固的；耐用的
shielded ['ʃiːldɪd] adj. 隔离的；屏蔽了的；防护的；铠装的
shipbuilding ['ʃɪpˌbɪldɪŋ] n. 造船；造船业；造船术
swirl [swɜːl] n. 漩涡；打旋；涡状形　vi. 盘绕；打旋；眩晕　vt. 使成漩涡
tubular ['tjuːbjʊlə] adj. 管状的
ventilating ['ventɪleɪtɪŋ] n. 通风　vt. 使通风（ventilate 的现在分词）

Exercises

1. Translate the following Chinese phrases into English

(1) 母体金属   (2) 熔焊   (3) 人工成本
(4) 表面防护   (5) 沉积速率   (6) 金属液滴
(7) 填充材料   (8) 低合金钢   (9) 真空系统
(10) 真空技术   (11) 高强度电流   (12) 低强度材料
(13) 爆炸焊   (14) 扩散接合   (15) 磁场

2. Translate the following English phrases into Chinese

(1) welding temperature   (2) electric arc
(3) equipment availability   (4) inert gas
(5) horizontal-vertical weld   (6) solidified slag
(7) contact tube   (8) high-energy heat source
(9) electron beam   (10) spot welding
(11) thermal damage   (12) rotational motion
(13) bimetallic structure   (14) prolonged cycle tine

3. Translate the following Chinese sentences into English

(1) 焊接是指在适当的温度、压力和冶金条件下,通过局部融合的方式将两块材料(通常是金属)永久连接在一起的过程。

(2) 金属极电弧焊利用电极和工件之间的电弧所提供的大约5 980 ℃的高温将工件的结合面熔化。

(3) 因此,该处对于焊接质量的控制就成为很重要的位置。

(4) 在焊接金属时所产生的烟雾中,常常包含了少量的锰。

(5) 焊接温度是焊管生产过程的重要参数。

(6) 数值模拟是研究电阻点焊过程和机理的重要手段。

(7) 为了提高镁合金耐腐蚀性能,对其进行表面防护处理是非常必要的。

4. Translate the following English sentences into Chinese

(1) The main advantages of the process are claimed to be increased Weld speed, deposition rate and penetration, reduced porosity and better tolerance to variations in fit-up compared with conventional single wire MIG/MAG welding.

(2) Oxyacetylene welding (OAW) is a process by which coalescence is achieved by heating a workpiece with flame produced by the combination of oxygen and acetylene gases.

(3) In summary, in order to obtain satisfactory welds, it is desirable to have a satisfactory heat and/or pressure source, a means of protecting or cleaning the metal, and avoidance of, or compensation for, harmful metallurgical effects.

(4) Autogenous welds are commonly used to evaluate corrosion rates of welded

materials and the usage of these materials in corrosive environments.

(5) It can detect flaws that could lead to cracks by fatigue, thermal damage and corrosion.

5. Translate the following Chinese essay into English

焊接分为两类：熔焊和塑焊或压力焊。在熔焊中，熔融金属自然凝固，而在压力焊中，熔融金属在有限空间内在压力下凝固，或者半固态金属在压力下冷却凝固。

熔焊细分成以下几类：

(1) 气焊

(2) 电弧焊，包括：

- 自动保护金属极电弧焊或手工电弧焊。
- 金属惰性气体电弧焊。
- 惰性气体保护钨极电弧焊。

(3) 铝热焊

压力焊或塑焊也细分成以下几类：

(1) 锻焊

(2) 电阻焊

6. Translate the following English essay into Chinese

Shielded metal arc welding (SMAW) is frequently referred to as "stick" or "covered electrode" welding. Stick welding is among the most widely used welding processes.

The flux covering on the electrode melts during welding. This forms the gas and slag to shield the arc and molten weld pool. The slag must be chipped off the weld bead after welding. The flux also provides a method of adding scavengers, deoxidizers, and alloying elements to the weld metal.

Whenan arc is struck between the metal rod (electrode) and the workpiece, both the rod and workpiece surface melt to form a weld pool. Simultaneous melting of the coating on the rod will form gas and slag which protects the weld pool from the surrounding atmosphere. The slag will solidify and cool, and must be chipped off the weld bead once the weld run is complete (or before the next weld pass is deposited).

扫一扫，查看更多资料

Chapter 9　Heat Treatment

9.1　Introduction

The heat treatment is a method of heating and cooling solidified steel for the purpose of improving some of its physical properties. The heat treating methods are based upon the iron-carbon equilibrium diagram. This part discusses how to use the eutectoid reaction, to control the structure and properties of steel through heat treatment. Four simple heat treatments process annealing, annealing, normalizing and spheroidizing are commonly used for steel. The heat treatment is used to accomplish one of three purposes: eliminating the effects of cold work, controlling dispersion strengthening, or improving machinability. Quenching hardens most steels and tempering increases the toughness. During tempering, an intimate mixture of ferrite and cementite forms from the martensite, the tempering treatment controls the final properties of the steel.

9.2　Full Annealing and Homogenizing

The term annealing has been used in its broadest sense to refer to any heat treatment that has as its objective the development of a nonmartensitic microstructure of low hardness and high ductility. This understanding of annealing is much too broad, however, and a number of more specific annealing heat treatments have been developed and defined. Full annealing[①] is a heat treatment accomplished by heating steels into the single phase-austenite field and slowly cooling, usually in a furnace, through the critical transformation ranges. When the term annealing is used without an adjective in reference to carbon steels, full annealing is the implied heat treatment practice.

Figure 9.1 shows the temperature ranges for several heat treatments involving austenitizing superimposed on the Fe－C diagram. As shown, the temperature for full annealing is a function of the carbon content of the steel, staying just above the A_3 temperature for hypoeutectoid steels and above the A_1 for hypereutectoid steels. The critical temperatures will vary somewhat with the alloy content of the steel, but the objective of heating into the single phase austenite field for low-carbon and medium-carbon steels and into the austenite-cementite field for high-carbon steels remains the same no matter what the steel composition is.

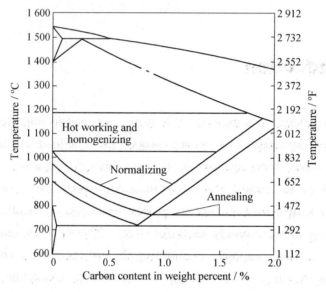

Figure 9.1 Portion of the Fe-C diagram with temperature ranges for full annealing, normalizing, hot working and homogenizing indicated

The reason for heating the hypereutectoid steels in the two-phase field is to agglomerate or spheroidize the proeutectoid cementite. If such steels are heated above A_{cm}, proeutectoid cementite would form on slow cooling at the austenite grain boundaries. The resulting network of carbides on the austenitic grain boundaries provides an easy fracture path and renders the steel brittle to forming or service stresses. Figure 9.2a shows carbide network developed in SAE 52 100 steel, a high-carbon bearing steel containing nominally 1 % carbon and 1.5 % chromium. Figure 9.2b shows how fracture produced by impact loading has followed the carbide network along prior austenite grain boundaries in a microstructure similar to that shown in Figure 9.2a. In Figure 9.2b, the steel has been hardened by quenching from the austenite-cementite field and martensite coexists with the carbide network. A carbide network formed on slow cooling from above A_{cm} in 52 100 steel is shown in Figure 9.3a. Pearlite instead of martensite has formed within the austenite grains. The object of full annealing high-carbon steels in the austenite-carbide field, then, is to break up such continuous carbide networks by agglomeration into separated spherical carbide particles. The driving force for this process is the reduction in austenite/cementite interface area and thus the reduction in interfacial energy that accompanies spheroidization. Figure 9.3b shows the partial spheroidization of a cementite network. Although the structure was formed during austenitizing for hardening, the austenitizing temperature ranges for hardening and full annealing are identical in high-carbon steels.

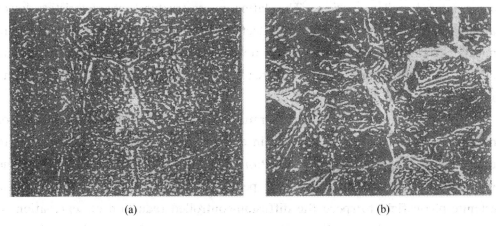

(a) (b)

Figure 9.2 Microstructure of SAE 52 100 steel

((a) —Carbide network at prior austenite grain boundaries in 52 100 steel (light micrograph, Nital ecth, magnification: 600×, shown here at 75 %). (b) —Fracture along grain boundary carbides in 52 100 steel (scanning electron micrograph; magnification: 415×, shown here at 75 %))

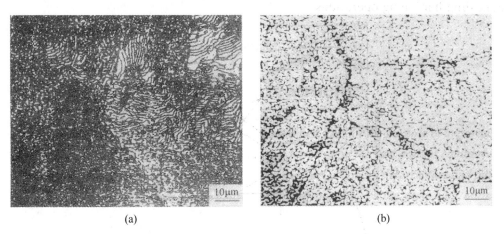

(a) (b)

Figure 9.3 Microstructure of SAE 52 100 steel

((a) — Proeutectoid cementite network in normalized 52 100 steel; (b) — Residual cementite network after austenitizing structure in Figure 9.3a at 850 ℃ (1 562 ℉) for hardening)

Not only the temperature range of heating is an important part of full annealing, but the slow cooling rate associated with full annealing is also a vital part of the process. Figure 9.4 compares schematic temperature-time schedules for full annealing and the normalizing heat treatments discussed in the next section of this part. The cooling rates are superimposed on a schematic CT diagram for hypoeutectoid steel. The slow cooling rates characteristic of furnace cooling insure that the austenite transforms first to proeutectoid ferrite and then to pearlite at temperatures approaching the equilibrium A_3 and A_1 temperatures. As a result, the ferrite will be equiaxed and relatively coarse-grained, and the pearlite will have a coarse inter lamellar spacing. The latter microstructural characteristics lower hardness and strength and increase ductility the major objectives of the full annealing treatment. Once the austenite has fully transformed to ferrite and pearlite, the cooling rate could be increased to reduce the time of annealing and thereby improve productivity.

Figure 9.4 also shows the temperature range for homogenizing, a type of annealing treatment usually performed in earlier stages of steel processing prior to hot rolling or forging, working operations that are also performed in the same temperature range. Homogenizing is performed at high temperatures in the austenite phase field to speed the diffusion-controlled reduction of segregation or chemical concentration gradients that are produced by ingot solidification. In addition, second phases such as carbides are dissolved as fully as possible. The resulting uniformity or homogeneity of the austenite not only improves hot workability, but also contributes to uniformity in the response of steel to subsequent annealing hardening operations.

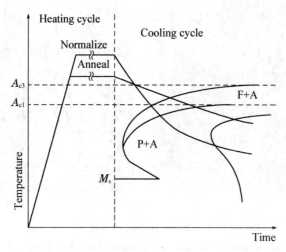

Figure 9.4 Schematic time—temperature cycles for normalizing and full annealing
(The slower cooling of annealing results in higher temperature transformation to ferrite and pearlite and coarser microstructures than dose normalizing)

9.3 Normalizing and Spheroidizing

9.3.1 Normalizing

Normalizing is a heat treatment which is similar to full annealing, produces a uniform microstructure of ferrite and pearlite. There are, however, several important differences between normalizing and annealing. Normalizing in hypoeutectoid steels is performed at temperatures somewhat higher than those used for annealing while in hypereutectoid steels the heating temperature range is above the A_{cm} (see Figure 9.1). In normalizing, heating is followed by air cooling, in contrast to the slower furnace cooling of full annealing.

The somewhat higher austenitizing temperatures used for normalizing, as compared to those used for annealing hypoeutectoid steels, in effect produce greater uniformity in austenitic structure and composition similar to a homogenizing treatment, although at a much lower temperature and for shorter time than those used for homogenizing. Another of the major objectives of normalizing is to refine the grain size that frequently becomes very coarse during hot working at high temperatures or that is present in as-solidified steel castings. As such hot worked or cast products are heated through the A_{c1} and A_{c3} temperatures, new austenite grains are nucleated and, if the austenitizing temperature is limited to the range shown in Figure 9.1, a uniform, fine-grained austenitic structure is produced. Normalizing, then, produces a uniform, fine-grained austenite grain structure that in hypoeutectoid steels transforms to ferrite-pearlite microstructures on air cooling. The resulting microstructure may have good uniformity and desirable mechanical properties for a given application or may be reaustenitized for final hardening by quenching to martensite.

In hypereutectoid steels, normalizing is performed above the A_{cm} not only to refine austenitic grain size but also to dissolve carbides and carbide networks that may have developed during prior processing. The normalized structures that result respond more readily to the spheroidizing treatments for good machinability as described below and/or provide better response to a subsequent and final hardening heat treatment. There is the possibility that continuous carbide networks may develop on cooling from a normalizing temperature above A_{cm}, and that as a result a somewhat brittle normalized microstructure might develop. On subsequent austenitizing for hardening, however, the carbides network agglomerate or spheroidize somewhat, and fracture toughness is in fact improved relative to a microstructure without the partially spheroidized network.

The air-cooling step of a normalizing treatment produces subtle but significant differences in microstructures compared to those produced by full annealing. Figure 9.4 shows schematically that air cooling lowers the temperature range over which proeutectoid ferrite and pearlite form compared to the transformation range in full annealing. As a result, both the ferrite grain size and the pearlite inter-lamellar spacing are reduced compared to those in the same steel in the fully annealed condition. The finer microstructure of the normalized steel in turn has higher strength and hardness and slightly lower ductility than the fully annealed steel.

The actual mechanical properties of any normalized or annealed steel are determined by a number of factors, the most important being carbon content. The higher the carbon content, the more pearlite that forms, and the higher the strength and hardness of the steel. Quantitative relationships for the contributions of carbon and other parameters to the mechanical properties of ferrite-pearlite steels are discussed in a later section of this part.

It is also important to realize that the air-cooling associated with a normalizing heat treatment produces a range of cooling rates depending on the section size. Heavier sections air-cool at much lower rates than do light sections because of the added time required for thermal conductivity to lower the temperature of central portions of the workpiece. Two important consequences follow from the effect of section size on cooling rate. In very heavy sections, the surface may cool at significantly higher rates than the interior, thus producing residual stresses. In very light sections, especially in alloy hardenable steels, air cooling may actually be rapid enough to produce bainitic or martensitic microstructures instead of ferrite and pearlite. The British Steel Corporation atlas which plots cooling transformation as a function of air-cooling section size enables this effect to be evaluated.

9.3.2 Spheroidizing

The most ductile, softest condition of any steel is associated with a microstructure that consists of spherical carbide particles uniformly dispersed in a ferrite matrix. Figure 9.5 shows a spheroidized microstructure of 0.66C-1Mn steel. The high ductility of such a microstructure is directly related to the continuous ductile ferrite matrix, pearlite with its fine lamellar carbides separating the ferrite, more effectively hinders deformation and, therefore, increases hardness and lowers ductility compared to a spheroidized structure. The good ductility of spheroidized microstructures is extremely important for low-carbon and medium-carbon steels that are cold formed, and the low hardness of spheroidized structures is important for high-carbon steels that undergo extensive machining prior to final hardening.

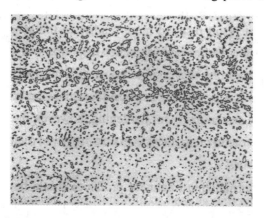

Figure 9.5　Spheroidized microstructure of Fe - 0.66C-1Mn alloy formed by heating martensite at 704 ℃ (1 300 ℉) for 24 h

Spheroidized microstructures are the most stable microstructures found in steels and will form in any prior structure, heated at temperatures high enough and time long enough to permit the diffusion dependent development of the spherical carbide particles. As a result, there are many different heat treatment approaches for producing spheroidized microstructures. The slowest spheroidizing is associated with pearlitic microstructures, especially those with coarse interlamellar spacing. Figure

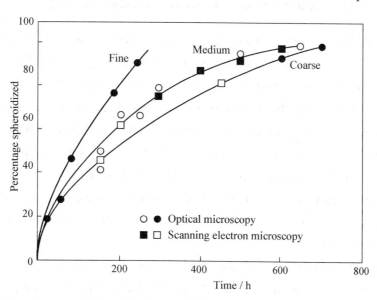

Figure 9.6　Progress of spheroidization at 700 ℃ (1 292 ℉) of fine medium and coarse pearlites in a steel containing 0.74 % C and 0.71 % Si

9.6 shows the percent of carbides that have spheroidized in fine to coarse pearlites produced by isothermally transforming an 0.074C – 0.71Si steel between 700 and 580 ℃ (1 292 and 1076 ℉), followed by annealing at 700 ℃ (1 292 ℉). Many hundreds of hours are required to spheroidize the pearlitic microstructures. Spheroidizing is more rapid if the carbides are initially in the form of discrete particles, as in bainite, and even more rapid if the starting structure is martensite. Spheroidizing of martensitic microstructures is most frequently performed on highly alloyed tool steels that form martensite on air cooling as shown schematically in Figure 9.7.

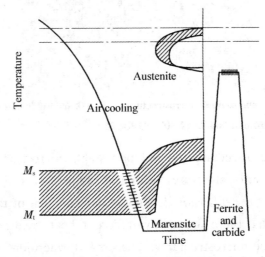

Figure 9.7 Schematic heat treatment cycle for spheroidizing an air-hardening steel (Martensite forms and then is tempered close to the A_{c1} to produce a spheroidize structure)

Spheroidizing at rates much faster than those shown in Figure 9.6 is accomplished by either complete or partial austenitizing, and then holding just below A_{c1}, cooling very slowly through the A_{c1}, or cycling above and below A_{c1}. These temperature ranges for spheroidizing are shown in Figure 9.8. It is important to limit the austenitizing temperature in order to retain a degree of heterogeneity in the austenite, especially since undissolved carbide particles appear to promote the transformation of the austenite to spheroidized microstructures. As noted earlier, homogenized austenite free of undissolved carbides as produced by normalizing or full annealing promotes the formation of pearlitic structures rather than spheroidized structures.

Spheroidized microstructures are stable because the ferrite is generally strain-free and because the spherical shape of the cementite particles is one of minimum interfacial area per unit volume of particle. Lamellar cementite particles, as present in pearlite, have a very large interfacial area per unit volume of particle and therefore high interfacial energy. In order to reduce the interfacial energy, cementite lamellae or plates are broken up into smaller particles that eventually assume spherical shapes. Figure 9.9 shows a representation of the breakup process of a single plate as determined by serial sectioning of a specimen annealed for 150 h at 700 ℃ (1 292 ℉). Once the lamellae have broken up, the small spherical particles dissolve at the expense of the larger particles, again driven by the reduction in interfacial energy.

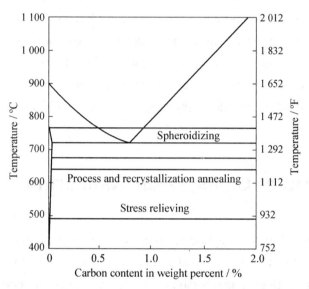

Figure 9.8　Portion of the Fe—C diagram with temperature ranges for process annealing, rectystallization annealing, stress relieving and spheroidizing indicated

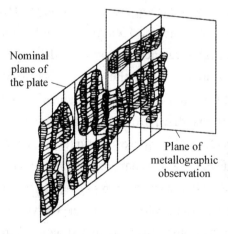

Figure 9.9　Representation of a partially spheroidized cementite plate in coarse pearlite structure annealed for 150 h at 700 ℃ (1 292 ℉)

The following equation describes the rate of coarsening of a spheroidized microstructure:

$$\frac{dr}{dt} = \frac{2rV_{Fe_3C}^2 \cdot X_C D_C^{eff}}{V_{Fe}RTr_1}\left(\frac{1}{\bar{r}} - \frac{1}{r_1}\right) \qquad (9-1)$$

where, r is the interfacial energy; $V_{Fe_3C}^{eff}$ and V_{Fe} are the molar volumes of cementite and ferrite; X_C is the mole fraction of carbon in equilibrium with cementite in ferrite; D_C^{eff} is the effective carbon diffusion coefficient; R is the gas constant; T is the absolute temperature; r_1 is the radius of newly created particles; $\bar{r}$ is the mean size of the already spheroidized particles. Equation shows that the rate of spheroidization is directly related to the diffusion of carbon in ferrite and decreases as the average size of particles in a spheroidized microstructure increases. Alloying elements slow the rate of carbon diffusion in ferrite, and therefore the spheroidization process. Also if present, strong carbide-forming elements would have to diffuse for alloy carbide coarsening, and thus would greatly reduce the rate of spheroidization.

9.4 Structural Changes on Tempering

The structure of steel quenched to form martensite is highly unstable. Reasons for the instability include the supersaturation of carbon atoms in the body-centered tetragonal crystal lattice of martensite, the strain energy associated with the fine dislocation or twin structure of the martensite, the interfacial energy associated with the high density of lath or plate boundaries, and the retained austenite that is invariably present even in low-carbon steels. The supersaturation of carbon atoms provides the driving force for carbide formation; the high strain energy the driving force for recovery; the high interfacial energy the driving force for grain growth or coarsening of the ferrite matrix; and the unstable austenite the driving force for transformation to mixtures of ferrite and cementite on tempering. Thus, even without the alloying effects discussed in the preceding section, there are many factors at work to produce the microstructures responsible for the mechanical property changes that develop when martensitic carbon steel is tempered.

An important series of papers on tempering carbon steels was published by Cohen and his colleagues in the 1950s. As a result of systematic X-ray, dilatometric and microstructural observations, three distinct stages of tempering were identified:

(1) Stage I: the formation of a transition carbide, epsilon carbide (or eta carbide as discussed below), and the lowering of the carbon content of the matrix martensite to about 0.25 % carbon.

(2) Stage II: the transformation of retained austenite to ferrite and cementite.

(3) Stage III: the replacement of the transition carbide and low-carbon martensite by cementite and ferrite.

The temperature ranges for the three stages overlap depending on the tempering time used, but the temperature ranges of 100 to 250 ℃ (212 to 482 ℉), 200 to 300 ℃ (392 to 572 ℉) and 250 to 350 ℃ (482 to 662 ℉) are generally accepted for the first, second and beginning third stages, respectively. The formation of the alloy carbides responsible for secondary hardening is sometimes referred to as the fourth stage of tempering. Also it is now recognized that carbon atom segregation to dislocations and various boundaries may occur during quenching and/or holding at room temperature and carbon atom clustering in as-quenched martensite may precede carbide formation that occurs in the first stage of tempering. Thus, tempering involves much more than three stages of tempering, but because of their central importance to understanding the behavior of tempered steels, the three stages listed above will be discussed in more detail.

The transition carbide that forms in the first stage of tempering was first identified as having a hexagonal structure and designated epsilon (ε) carbide by Jack. More recently, Hirotsu and Nagakurea have shown that the transition carbide has an orthorhombic structure isomorphous with transition metal carbides of the M_2C type. The transition carbide with the latter structure was designated as eta (η) carbide. The structures of the epsilon and eta carbide are very similar and are differentiated primarily by electron diffraction spots that come from a regular array of carbon atoms (or sublattice of carbon atoms) in the eta carbide. Both the epsilon carbide, $Fe_{2.4}C$, and the eta carbide, Fe_2C, have carbon contents substantially higher than that of the cementite, Fe_3C, which forms at higher temperatures. Kinetic studies show that the first stage of tempering is dependent on the diffusion of carbon through the martensite with activation energy of 16 000 cal/mol (1 cal = 4.186 8 J).

Figure 9.10 to Figure 9.12 are transmission electron micrographs that show various aspects of the transition carbide formation in the martensite of a Fe-1.22C alloy tempered at 150 ℃ (302 ℉) for 16 h. Figure 9.10 shows a typical plate martensitic microstructure with plates of a variety of sizes and patches of retained austenite (black areas) between the plates. Each of the plates contains a highly uniform distribution of fine carbide particles. Figure 9.11 shows a typical array of transition carbides, identified as eta carbides, in a single plate of martensite. The carbides appear to be in the form of fine platelets, but Figure 9.12, a dark field micrograph taken with illumination from a carbide diffraction spot shows that the eta carbide is actually present as rows of fine spherical particles about 2 nm in diameter. The dark contrast in the plate-like morphology associated with the carbides in Figure 9.11, is apparently due to strain effects between the martensitic matrix and the rows of particles.

Figure 9.10 Transmission electron micrograph of martensitic microstructure in a Fe-I.2C alloy tempered of 150 ℃ (302 ℉)
(The microstructure consists of plates of various sizes containing uniform arrays of very fine carbides and retained austenite (black patches))

Figure 9.11 Transmission electron micrograph of distribution of eta carbide in martensite plate of a Fe-1.22C alloy tempered at 150 ℃ (302 ℉) for 16 h
(Magnification: 80 000×, shown here at 75 %)

The transformation of retained austenite during tempering occurs only after the transition carbide is well established. Figure 9.13 shows the rate of transformation of the retained austenite in a Fe - 1.22C alloy at three different tempering temperatures. About 19 % retained austenite, distributed as shown in Figure 9.10, was initially present in the as-quenched structure. Even at 180 ℃ (356 ℉) the retained austenite transformed completely to mixtures of ferrite and cementite if held for sufficiently long time. Analysis of the austenite transformation kinetics in

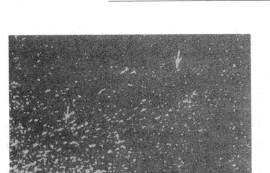

Figure9.12 Dark-field transmssion electron micrograph of row of fine spherical eta carbide particles in a martensite plate of a Fe—1.22C alloy tempered of 150 ℃ (302 °F) for 16 h (Magnification: 80 000×, shown here at 75 %)

Figure 9.13 yielded activation energy of 1.15×10^5 J/mol (27 kcal/mol) in good agreement with the activation energies for the diffusion of carbon in austenite and the activation energy for the second stage of tempering reported by Roberts, Averbach and Cohen. Figure 9.14 shows that retained austenite is present in small amounts, about 2 % and 4 %, in as-quenched specimens of 4 130 and 4 340 steels, respectively, and that for tempering time of 1 h, the transformation of retained austenite in these low-alloy medium-carbon steels begins only above 200 ℃ (392 °F). Transformation is complete at about 300 ℃ (572 °F) and cementite becomes an important part of the microstructure, after tempering at 300 ℃ (572 °F) and higher temperatures.

The third stage of tempering consists of the formation of ferrite and cementite as required by the Fe－C diagram. However, there is some evidence, especially in high-carbon steels, that Hagg or chi (X) carbide formation precedes cementite or theta (θ) carbide formation. The chi carbide has a monoclinic structure, and the composition Fe_5C_2. However, despite the differences between cementite and chi carbide, the relatively complex structures of the two carbide phases are similar and difficult to separate by X-ray or electron diffraction techniques. Therefore, in view of the experimental difficulty in separating the presence of chi carbide from that of cementite, the temperature and compositions of the steels in which chi carbide forms are not yet completely defined.

Chapter 9 Heat Treatment

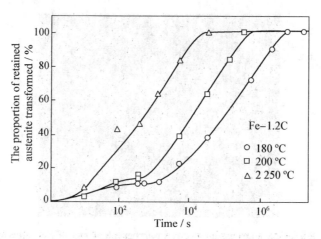

Figure 9.13 Transformation of retained austenite in a Fe—1.22C alloy as a function of time at three tempering temperature

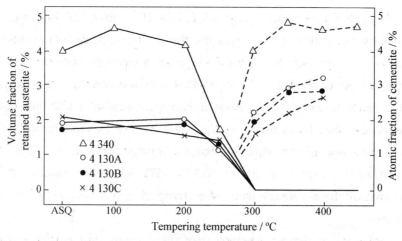

Figure 9.14 Retained austenite and cementite as a function of tempering temperature in 4 340 and 4 130 type steels

(The amounts of the phases were determined by Mossbauer spectroscopy)

Figure 9.15 shows the dense carbide distribution that has formed in the martensite of a Fe − 1.22C alloy tempered at 350 ℃ (663 ℉). In this case, the carbides were best identified as chi carbide. Two carbide morphologies are present: those that have nucleated and grown within the martensite plates, and very long planar carbides that have formed along the plate interfaces, perhaps as a result of the transformation of retained austenite in the second stage of tempering. A third morphology of chi carbide and/or cementite in tempered high-carbon steels consists of parallel arrays of carbides formed on transformation, twins sometimes present in high-carbon martensite, especially in the midrib portions of the plates. The carbides that have formed within the plates are coarser than the transition carbides and will eventually spheroidize if tempering is performed, at higher temperatures.

Figure 9.15 Transmission electron micrograph of cementite and orchid-carbide formation in martensitic structure of a Fe—1.22C alloy tempered of 350 ℃ (622 ℉) for 1 h
(Magnification: 30 000×, shown here at 75 %)

The carbide structures and distributions that form in alloy steels and retard softening and/or produce secondary hardening during tempering are quite varied. Many of the alloy carbides and their formation on tempering have been characterized by Honeycombe and his colleagues. Much of this work includes descriptions of the carbide structures produced by tempering vanadium, molybdenum, tungsten, chromium and titanium steels. The alloy carbide distributions formed in the secondary hardening range, 500 to 650 ℃ (932 to 1 202 ℉), depend on the nature of the cementite distribution formed at lower tempering temperatures, and the nature of the transformation of cementite to the alloy carbide. Honeycombe presents evidence for two basic modes of alloy carbide formation on tempering. The carbides may form directly from the cementite, a mode referred to as in situ transformation, or the carbides may form by separate nucleation, after the cementite particles dissolve in the ferrite matrix. The independently nucleated alloy carbide particles are often nucleated on the dislocations residual from the as-quenched martensite, and tend to be much finer than the alloy carbides nucleated on the cementite particles. Most of the structural changes discussed above have involved the formation of various types of carbides during tempering. There are also important changes in the martensitic matrix that accomplish the formation of fully tempered structures consisting of spheroidized carbides in a matrix of equiaxed ferrite grains. Figure 9.16 to Figure 9.19 show changes in the matrix structure that developed during the tempering of lath martensite in a Fe−0.2C alloy. Figure 9.16 shows that tempering at 400 ℃ (752 ℉) for 15 min produces little change from the appearance of as-quenched lath martensite on the scale resolvable with the light microscope. More pronounced changes are visible in a specimen tempered at 700 ℃ (1 292 ℉) (see Figure 9.17), but even after this rather severe temper, the packet morphology with its parallel sub-units is still clearly visible. The major effects of tempering have been to eliminate many of the smaller laths and to produce coarse, spherical cementite particles at the prior austenite grain boundaries and within the packets. More severe tempering, 700 ℃ (1 992 ℉) for 12 h, begins to break up the remaining parallel blocks of crystals within the packets and more equiaxed ferrite grains begin to form (see Figure 9.18), the equiaxed grains contain subboundaries made up of regular dislocation arrays as shown in the electron micrographs of Figure 9.19.

Figure 9.16 Microstructure of lath martensite in a Fe－0.2C alloy after tempering at 400 ℃ (752 ℉) for 15 min
(Light micrograph. Nital etch. Magnification: 500×)

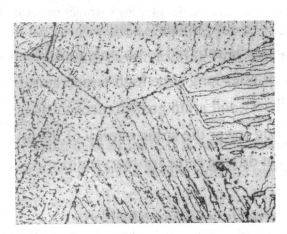

Figure 9.17 Microstructure of lath martensite in a Fe－0.2C alloy after tempering at 700 ℃ (1 292 ℉) for 2 h
(Light micrograph. Nital etch. Magnification: 500×)

Figure 9.18 Microstructure of lath martensite in a Fe—0.2C alloy after tempering at 700 ℃ (1 292 ℉) for 12 h

(Light micrograph, Nital etch, Magnification: 500×)

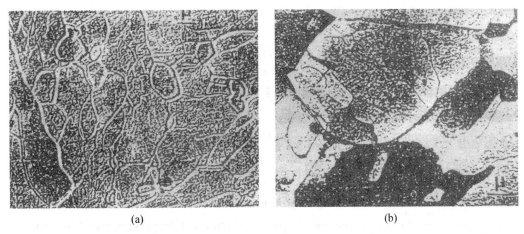

(a)　　　　　　　　　　　　　　(b)

Figure 9.19 Structure of lath martensite in a Fe—0.2C alloy after tempering at 700 ℃ (1 292 ℉) for 12 h

(a) Single stage replica of polished and etched (natal) surface; (b) Transmission electron micrograph

Systematic measurement of the change in lath boundary per unit volume as a function of tempering of the Fe − 0.2C martensite shows that the very high lath boundary area per unit volume of the fine laths in as-quenched martensite decreased very rapidly on tempering. This initial rapid decrease is primarily due to the elimination of the low-angle boundaries between laths of similar orientation.

Simultaneously, fine carbides precipitate and help to stabilize the surviving lath boundaries to maintain their parallel orientation within the packets. All of these initial matrix changes occur as a result of recovery mechanisms. The dislocation density is effectively lowered not only by the reduction of dislocations within the laths but also by the elimination of the low-angle lath boundaries. Eventually, with coarsening of the carbide particles, the remaining large-angle boundaries rearrange themselves to produce more equilibrium junctions between grains as typical of the mechanisms associated with grain growth. Any residual dislocations within the laths then rearrange themselves into low-angle boundaries within the equiaxed grains as shown in Figure 9.19. Such subdivision of large grains by dislocation boundaries is referred to as polygonization. Thus, the formation of the equiaxed ferrite matrix that develops after long-time and high-temperature tempering of a low-carbon lath martensite is accomplished by recovery and grain growth mechanisms. Apparently, the recovery mechanisms that operate early in tempering lower the strain energy of the as-quenched martensite to the point where there is no longer sufficient driving force for recrystallization.

9.5 Thermomechanical Treatments

Thermomechanical treatments[2] are processing treatments that combine plastic deformation with thermal processing or heat treatment in order to produce microstructures and improved properties not obtained by independently applied conventional heat treatment or working operations. Generally, increased strength with improvement in ductility and/or toughness is the objectives of thermomechanical treatments. Intensive research in the 1950s and 1960s showed that these objectives could be achieved, and resulted in a U.S. classification system that is recognized three types of thermomechanical treatments. Table 9.1, after Azrin, describes the three classes of thermomechanical treatment and Figure 9.20 shows the treatments schematically superimposed on an idealized transformation diagram. The Soviets have also actively pursued the development of thermomechanical treatments and their classification systems shown in Table 9.2 and Figure 9.21, also after Azrin. As Figure 9.20 shows, the thermomechanical treatments, recognized as such, have the modification of martensite (either by the deformation of the austenite preceding transformation or by the deformation of the martensite after transformation) as the major approach to improve properties. Very high strengths therefore result from this type of thermomechanical process.

Table 9.1 U.S. classification of thermomechanical treatments

CLASS I	Deformation occurs before the austenite transformation. Austenite is deformed in the stable austenite range above the critical temperature (A_1) or in the unstable region above the pearlite nose or in the bay region between the pearlite and bainite noses.

(to be continued)

CLASS II	Deformation during the austenite transformation. Depending on the deformation temperature, as well as the M_s and M_D temperatures, the transformation products can be either pearlite, bainite or martensite. The martensite transformation can be either pearlite, bainite or martensite. The martensite transformation can be due to a strain-induced or stress-assisted transformation.
CLASS III	Deformation after austenite transform to martensite or other transformation products. M_D is temperature above which martensite can not be formed by plastic deformation.

Table 9.2 Soviet classification of thermomechanical treatments

Soriet classification	Thermomechanical treatments
SHT	Standard heat treatment-conventional heat treatment without deformation
TMT	Thermomechanical treatment-a combined thermal and mechanical treatment generally involving a phase transformation
HTTMT	High temperature thermomechanical treatment-deformation above the recrystallization temperature
LTTMT	Low temperature thermomechanical treatment-deformation below the recrystallization temperature
CTMT	Combined thermomechanical treatment-HTTMT followed by LTTMT
PTMT	Preliminary thermomechanical treatment-deformation by HTTMT or LTTMT or cold working followed by rapid reaustenitizing and quenching
MTT	Mechanico-thermal treatment-deformation at room or elevated temperature with or without subsequent annealing or aging applied to a material which does not undergo a phase transformation. As with TMT, deformation can be below (LTMTT) or above (HTMTT) the recrystallization temperature

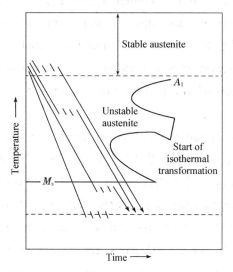

Figure 9.20 Schematic diagram of the U. S. classification of thermomechanical treatments superimposed on a time-temperature transformation diagram

Chapter 9 Heat Treatment

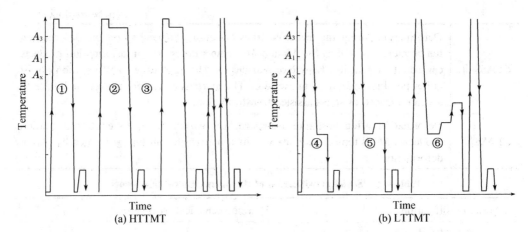

Figure 9.21 Schematic time-temperature deformation cycles for Soviet thermomechanical treatments (Treatment ② and ⑤ include polygonization, treatment ③ and ⑥ include hereditary treatments. R_x is the minimum recrystallization temperature)

Ausforming is a special term for the class I type of thermomechanical treatment in which austenite is deformed in the bay of the time-temperature transformation diagram prior to martensite formation. A special variant of this treatment involves the use of highly alloyed steels with an initial structure containing metastable austenite. These steels are referred to as transformation induced plasticity (TRIP) steels because they transform to martensite on straining. The strain induced martensite in turn resists plastic instability or necking and thereby produces good combinations of very high strength and ductility. Figure 9.21, after Zaokay et al., compares strengths and ductilities produced by various thermomechanical techniques. The properties of the low-alloy and high-strength steels indicated are those developed by conventional quench and tempering of carbon steels. The maraging steels are highly alloyed steels, containing about 18 % nickel, 8 % cobalt, and small amounts of aluminum and titanium. High strength and good toughness are obtained by precipitation of compounds such as Ni_3Mo and Ni_3Ti in a matrix of carbon-free lath martensite.

While the physical metallurgy of thermomechanical treatments such as those shown in Figure 9.20 and Figure 9.21 are well established, very few industrial applications of thermomechanical treatments have been developed. Reasons for this situation include technical factors such as the difficulty in machining the very hard thermomechanically treated products; the deterioration in mechanical properties when thermomechanically processed parts are joined by welding; the high-alloy content of steels that respond best to thermomechanical treatment, and the difficulty in achieving the uniform deformation required for improved properties in parts with complex shapes. The status of thermomechanical treatments in the U.S. and the U.S.S.R. is reviewed in the reports by Henning and Azrin, respectively, and the reader is referred to these publications for the practical aspects and limitations of thermomechanical treatments as well as extensive bibliographies of the literature on thermomechanical treatments.

Perhaps the most important application of thermomechanical processing actually in use in the controlled rolling of microalloyed HSLA steels in regard to grain size control. The processing of HSLA low-carbon steels, however, depends on very fine ferrite grain sizes and precipitation for strengthening and not martensite formation, and therefore falls outside of the classification systems shown in Table 9.1.

9.6 Surface Hardening

Surface hardening[3] is used to extend the versatility of certain steels by producing combinations of properties not readily attainable in other ways. For many applications, wear and the most severe stresses act only on the surface of a part. Therefore, the part may be fabricated from a formable low-carbon or medium-carbon steel, and is surface hardened by a final heat treatment after all other processing has been accomplished. Surface hardening also reduces distortion and eliminates cracking that might accompany through hardening, especially in large sections. Localized hardening of selected areas is also possible by means of certain surface hardening techniques. This part describes two major approaches to surface hardening. One approach does not change composition and consists of hardening the surface by flame or induction heating. The other approach changes the surface composition and includes the applications of such techniques as carburizing, nitriding and carbonitriding.

9.6.1 Flame Hardening

Flame hardening[4] consists of austenitizing the surface of steels by heating with an oxyacetylene or oxyhydrogen torch and immediately quenching with water. A hard surface layer of martensite over a softer interior core with a ferrite-pearlite structure results. There is no change in composition, and therefore the flame-hardened steel must have adequate carbon content for the desired surface hardness. The rate of heating and the conduction of heat into the interior appear to be more important in establishing case depth than having a steel of high hardenability. Figure 9.22 shows hardness gradients produced by various rates of flame travel across a 1 050 steel forging. The slower the rate of travel, the greater the heat penetration and the depth of hardening.

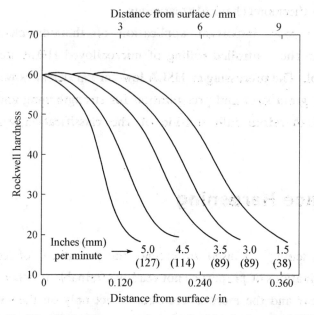

Figure 9.22 The effect of flame speed on depth of hardening of a 1 050 forging

A number of different methods of flame hardening have been developed. Localized or spot hardening may be performed by directing a stationary flame head to an area of a stationary workpiece. Progressive methods where the torch travels over the workpiece or the workpiece travels under a stationary torch and quenching fixture are used for long bars. Spinning methods in which the workpiece is rotated within an array of torches are often used for small rounds. In this method heating is performed first, then the flames are extinguished, and quenching is finally accomplished by water sprays or dropping the part into a quench tank. In all cases, the quenched parts are tempered to improve toughness and relieve stresses induced by the surface hardening.

9.6.2 Induction Heating

Induction heating[5] is an extremely versatile method for hardening steel. Uniform surface hardening, localized surface hardening, through hardening, and tempering of hardened pieces may all be performed by induction heating. Heating is accomplished by placing a steel part in the magnetic field generated by high-frequency alternating current passing through an inductor usually a water-cooled copper coil. The rapidly alternating magnetic field established within the coil induces current (I), within the steel. The induced currents then generate heat (H) according to the relationship $H = I^2 R$ where R is the electrical resistance. Steel, consisting primarily of ferrite or bcc iron, is ferromagnetic up to its Curie temperature[6] (768 ℃ or 1 400 ℉), and the rapid change in direction of the internal magnetization of domains in a steel within the field of the coil also generates considerable heat. When steel transforms to austenite, which is nonmagnetic, this contribution to induction heating becomes negligible. A wide variety of heating patterns may be established by induction heating depending on the shape of the coil, the number of turns of the coil, the operating frequency and the alternating current power input. Figure 9.23 shows examples of the heating patterns produced by various types of coils.

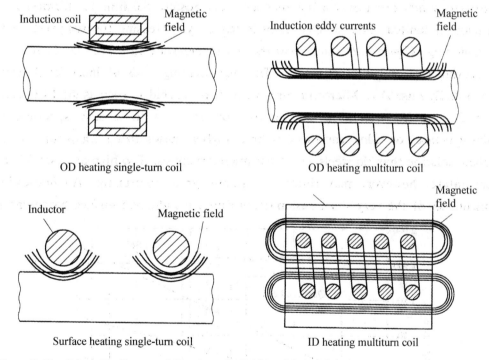

Figure 9.23 Schematic diagram of the magnetic fields and induced currents produced by several types of induction coils

The depth of heating produced by induction is related to the frequency of the alternating current. The higher the frequency, the thinner or more shallow the heating. Therefore, deeper case depths and even through hardening are produced by using lower frequencies.

As in flame hardening, induction heating does not change the composition of steel, and therefore steel selected for induction hardening must have sufficient carbon content and alloying for the desired surface hardness distribution. Generally, medium-carbon and high-carbon steels are selected because the high surface strengths and hardness attainable in these steels significantly improve fatigue and wear resistance. Induction hardening introduces residual compressive stresses into the surface of hardened parts. Therefore, the fatigue strengths of induction surface hardened parts may be higher than those of through hardened parts in which quenching develops residual surface tensile stresses that may only be partly relieved during tempering. The greater the depth of hardening by induction heating, however, the more the surface stress state approaches that of through hardening. Too deep a hardened case may in fact cause surface tensile stresses and even cracking of susceptible steels.

The duration of high-frequency induction heating cycles for surface hardening is extremely short, often only a few seconds. As a result, the time for formation of austenite is limited, and compensation is made by increasing the temperature of austenitizing. Figure 9.24 shows how the A_{c3} temperature in 1 042 steel is affected by heating rate and microstructure. The high heating rates of induction heating substantially raise A_{c3}. Microstructures with coarse carbides, such as the 1 042 steel in the annealed condition as shown, or steels with coarse spheroidized microstructures or alloy carbides, require higher austenitizing temperatures for carbide solution than do steels with finer microstructures. Too high an austenitizing temperature, however, may result in austenite grain coarsening. An interesting consequence of the very short austenitizing time for induction surface hardening is

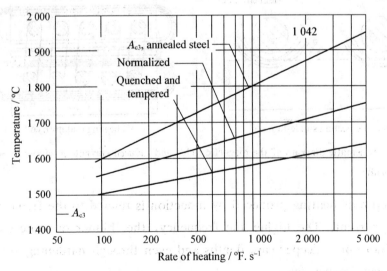

Figure 9.24 Change in A_{c3} temperature of 1 042 steel as a function of microstructure and heating rate

the development of the hardness above that normally expected for through-hardened martensite. This higher hardness is sometimes referred to as superhardness (see Figure 9.25) and may be a result of martensite formed in very fine grained, imperfect austenite produced by the short-time austenitizing treatments used in induction surface hardening.

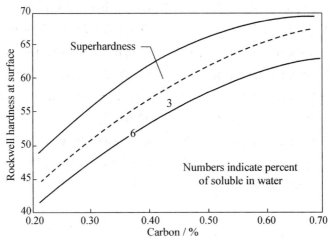

Figure 9.25 Superhardness produced by induction hardening compared to that produced by conventional furnace hardening (lower solid curve)

Notes

① full annealing 完全退火
② thermomechanical treatments 形变热处理
③ surface hardening 表面硬化
④ flame hardening 火焰硬化,淬火硬化
⑤ induction heating 感应加热
⑥ Curie temperature 居里点,居里温度

Vocabulary

agglomeration [əɡlɒməˈreɪʃ(ə)n] n. 凝聚;结块;附聚
annealing [əˈniːlɪŋ] 热处理;低温退火 v. [机][生化]退火
bainitic [beɪˈnɪtɪk] n. 贝氏体
carbonitriding [ˌkɑːbənˈnaɪtraɪdɪŋ] n. [材]碳氮共渗
carburizing [ˈkɑːbjʊraɪzɪŋ] n. [材]渗碳 v. 渗碳于(carburize 的 ing 形式)

dilatometric [dɪleɪˈtəʊmetrɪk] adj. 测膨胀的
dispersion [dɪˈspɜːʃ(ə)n] n. 散布;[统计][数]离差;驱散
hardenability [ˌhɑːdənəˈbɪlətɪ] n. 可硬化性;淬硬性
hypoeutectoid [ˌhaɪpəʊjʊˈtektɔɪd] n. 低易熔质 adj. 亚共析的
induction [ɪnˈdʌkʃ(ə)n] n. [电磁]感

Chapter 9 Heat Treatment

应;归纳法
lath [lɑːθ] n. 板条 vt. 给……钉板条
machinability [məˌʃiːnəˈbɪlətɪ] n. 切削性;机械加工性
martensite [ˈmɑːtɪnzaɪt] n. [材]马氏体
nitriding [ˈnaɪtraɪdɪŋ] n. 渗氮法 v. [材]渗氮(nitride 的 ing 形式)
normalizing [ˈnɔːməlaɪzɪŋ] n. [冶]正火 v. 对钢正火(normalize 的 ing 形式
oxyhydrogen [ˌɒksɪˈhaɪdrədʒən] n. 氢氧混合气 adj. 氢氧的;氢氧混合的
polygonization [ˌpɒlɪɡənaɪˈzeɪʃən] n. 多边化;[物]多边形化
recrystallization [riːˌkrɪstəlaɪˈzeɪʃən] n. 再结晶
spheroidization [sfɪərɒɪdaɪˈzeɪʃn] n. 球状化处理
supersaturation [ˈsjuːpəˌsætʃəˈreɪʃən] n. 过度饱和
susceptible [səˈseptɪb(ə)l] adj. 易受影响的;易感动的;容许……的
tempering [ˈtempərɪŋ] n. [机]回火 v. 调和(temper 的 ing 形式)
tetragonal [tɪˈtræɡ(ə)n(ə)l] adj. [数]四角形的
thermomechanical [ˌθɜːməʊmɪˈkænɪkəl] adj. [热]热机的,[热]热机械的

◆ Exercises

1. Translate the following Chinese phrases into English

(1) 热处理 (2) 完全退火 (3) 铁碳相图
(4) 中碳钢 (5) 冷却循环 (6) 空气硬化钢
(7) 气体常数 (8) 回火温度 (9) 形变热处理
(10) 表面硬化 (11) 感应加热 (12) 界面能
(13) 发热

2. Translate the following English phrases into Chinese

(1) eutectoid reaction (2) hypoeutectoid steel
(3) critical temperature (4) austenitic grain boundary
(5) spheroidized microstructure (6) diffusion coefficient
(7) ferrite matrix (8) equiaxed ferrite grain
(9) high-strength steel (10) flame-hardened steel
(11) high-frequency induction (12) critical temperature

3. Translate the following Chinese sentences into English

(1) 转变的终止以奥氏体的形成和珠光体的消失为标志。
(2) 正如用 X 线发和磁热分析法所揭示的那样,低温回火时形成的亚稳碳化物与渗碳体不同。
(3) 该浓度不均匀导致产生应力,而且因为贫碳区的马氏体相变点高于保温温度,塑性变形将诱发 γ 以马氏体转变机制向 α 的转变。
(4) 回火钢的特性使得我们可将其用于高硬工具的制造。
(5) 从铁碳相图可知,只有在缓慢加热时珠光体才能向奥氏体转变。

(6) 黑色金属(含钢的金属)通常的热处理方法为退火、正火、硬化以及回火。

4. Translate the following English sentences into Chinese

(1) Since these boundaries are very developed, the transformation starts from formation of a multitude of fine grains.

(2) The first transformation on tempering produces temper martensite which is a heterogeneous mixture of supersaturated α solution of an inhomogeneous concentration and nonisolated carbide particles.

(3) Whereas the described mechanism is valid for the whole temperature range of bainite transformation, a change of temperature within that range may cause appreciable quantitative differences.

(4) In the first place, steel which contains very little carbon will be milder than steel which contains a higher percentage of carbon, up to the limit of about 1.5%.

(5) The process of heat treatment is the method by which metals are heated and cooled in a series of specific operations that never allow the metal to reach the molten state.

5. Translate the following Chinese essay into English

热处理就是将金属加热到一定温度,并以特定的方式冷却来改变其内部结构获得期望的物理和机械性能,如脆性、硬度和柔性的工艺。普遍接受的热处理金属和合金的术语是以一种特定的方式来加热和冷却固体金属或合金,从而获得特殊的状态或性能。

6. Translate the following English essay into Chinese

In some instances, heat treatment procedures are clear-cut in terms of technique and application, whereas in other instances, descriptions or simple explanations are insufficient because the same technique may be used frequently to obtain different objectives. For example, stress relieving and tempering are often accomplished with the same equipment and by use of identical time and temperature cycles. The objectives, however, are different for the two processes. Heat treatment is divided into surface heat treatment and full heat treatment. The following introduction is about the four important heat treatment processes of full heat treatment.

Annealing is a generic term denoting a treatment consisting of heating to and holding at a suitable temperature, followed by cooling at a suitable rate. The process is used primarily to soften metal sand to simultaneously produce desired changes in other properties or in microstructures.

Normalizing is a homogenizing or grain refining treatment, with the aim of being uniformity in composition throughout a part. In the thermal sense, normalizing is an austenitizing heating cycle followed by cooling in still or slightly agitated air.

Stainless and high-alloy steels may be quenched to minimize the presence of grain-boundary carbides or to improve the ferrite distribution, but most steels, including carbon, low-alloy and tool steels, are quenched to produce controlled amounts of martensite in the microstructure.

In the process of tempering, previously hardened or normalized steel is usually heated to a temperature below the lower critical temperature and cooled at a suitable rate, primarily to increase ductility and toughness, but also to increase grain size of the matrix.

扫一扫,查看更多资料

Chapter 10 Ceramic Processing Methods

10.1 Introduction

Fabrication techniques are those methods by which materials are formed or manufactured into components that may be incorporated in useful products. Sometimes it also may be necessary to subject the component to some type of processing treatment in order to achieve the required properties. And, on occasion, the suitability of a material for an application is dictated by economic considerations with respect to fabrication and processing operations.

As with any materials process, the unit operations we evaluate are powder preparation, forming, consolidation and sintering. Some of the common steps in ceramic manufacture are shown in Figure 10.1.

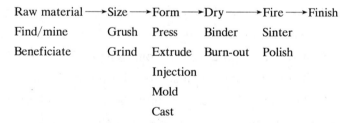

Figure 10.1 Unit operations in manufacturing ceramic part

10.2 Forming Processing

Some ceramic pieces are formed from powders (or particulate collections) that must ultimately be dried and fired. Glass shapes are formed at elevated temperatures from a fluid mass that becomes very viscous upon cooling. Cements are shaped by placing into forms a fluid paste that hardens and assumes a permanent set by virtue of chemical reactions. A taxonomical scheme for the several types of ceramic-forming techniques is presented in Figure 10.2.

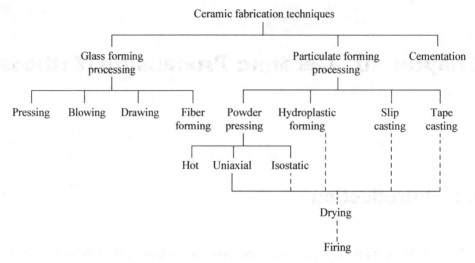

Figure 10.2　A classification scheme for the ceramic-forming techniques

10.3　Glass Forming Processing

Glass is produced by heating the raw materials to an elevated temperature above which melting occurs. Most commercial glasses are of the silica-soda-lime variety; the silica is usually supplied as common quartz sand, whereas Na_2O and CaO are added as soda ash (Na_2CO_3) and limestone ($CaCO_3$). For most applications, especially when optical transparency is important, it is essential that the glass product be homogeneous and pore free. Homogeneity is achieved by complete melting and mixing of the raw ingredients. Porosity results from small gas bubbles that are produced; these must be absorbed into the melt or otherwise eliminated, which requires proper adjustment of the viscosity of the molten material.

Four different forming methods are used to fabricate glass products: pressing, blowing, drawing and fiber forming. Pressing is used in the fabrication of relatively thick-walled pieces such as plates and dishes. The glass piece is formed by pressure application in a graphite-coated cast iron mold having the desired shape; the mold is ordinarily heated to ensure an even surface.

Although some glass blowing is done by hand, the process has been completely automated for the production of glass jars, bottles and light bulbs. A raw go of glass, a parson, temporary shape, is formed by mechanical pressing in a mold. This piece is inserted into a finishing or blow mold and forced to conform to the mold contours by the pressure created from a blast of air.

Drawing is used to form long glass pieces such as sheet, rod, tubing and fibers, which have a constant cross section. One process by which sheet glass is formed is illustrated in Figure 10.3; it may also be fabricated by hot rolling. Flatness and the surface finish may be improved significantly by floating the sheet on a bath of molten tin at an elevated temperature; the piece is slowly cooled and subsequently heat treated by continuous glass fibers are formed in a rather sophisticated drawing operation.

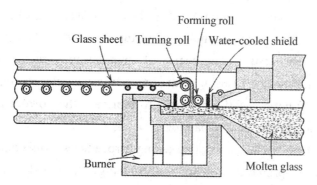

Figure 10.3 A process for the continuous drawing of sheet glass

10.4 Particulate Forming Processing

The raw materials usually have to go through a milling or grindiny operation in which particle size is reduced; this is followed by screening or sizing to yield a powdered product having a desired range of particle sizes. Four common shaping techniques are utilized: hydroplastic forming, slip casting, powder pressing and tape casting.

The most common hydroplastic forming technique is extrusion, in which a stiff plastic ceramic mass is forced through a die orifice having the desired cross-sectional geometry. Brick, pipe, ceramic blocks and tiles are all commonly fabricated using hydroplastic forming. Usually the plastic ceramic is forced through the die by means of a motor-driven auger[①], and often air is removed in a vacuum chamber to enhance the density.

The second forming process is slip casting. A slip is a suspension of clay and/or other nonplastic materials in water. When poured into a porous mold (commonly made of plaster), water from the slip is absorbed into the mold, leaving behind a solid layer on the mold wall, the thickness of which depends on the time. This process may be continued until the entire mold cavity becomes solid (solid casting).

Chapter 10 Ceramic Processing Methods

Or it may be terminated when the solid shell wall reaches the desired thickness, by inverting the mold and pouring out the excess slip; this is termed drain casting. As the cast piece dries and shrinks, it will pull away (or release) from the mold wall, at this time the mold may be disassembled and the cast piece removed.

Another important and commonly used method that warrants a brief treatment is powder pressing. Powder pressing, the ceramic analogue to powder metallurgy, is used to fabricate both clay and nonclay compositions, including electronic and magnetic ceramics as well as some refractory brick products. In essence, a powdered mass, usually containing a small amount of water or other binder, is compacted into the desired shape by pressure. The degree of compaction is maximized and fraction of void space is minimized by using coarse and fine particles mixed in appropriate proportions. One function of the binder is to lubricate the powder particles as they move past one another in the compaction process.

There are three basic powder pressing procedures: uniaxial, isostatic (or hydrostatic) and hot pressing. For uniaxial pressing, the powder is compacted in a metal die by pressure that is applied in a single direction. This method is confined to shapes that are relatively simple; however, production rates are high and the process is inexpensive. For isostatic pressing, the powdered material is contained in a rubber envelope and the pressure is applied by a fluid, isostatically (i. e., it has the same magnitude in all directions). More complicated shapes are possible than with uniaxial pressing; however, the isostatic technique is more time consuming and expensive. With hot pressing, the powder pressing and heat treatment are performed simultaneously, and the powder aggregate is compacted at an elevated temperature. The procedure is used for materials that do not form a liquid phase except at very high and impractical temperatures; in addition, it is utilized when high densities without appreciable grain growth are desired. This is an expensive fabrication technique that has some limitations.

Tape casting is a relatively new and important ceramic fabrication technique. As the name implies, thin sheets of a flexible tape are produced by means of a casting process. These sheets are prepared from slips, in many respects similar to those that are employed for slip casting. This type of slip consists of a suspension of ceramic particles in a liquid that also contains binders and plasticizers that are incorporated to impart strength and flexibility to the cast tape. Deairing in a vacuum may also be necessary to remove any entrapped air or solvent vapor bubbles, which may act as crack-initiation[2] sites in the finished piece. The actual tape is formed by pouring the slip onto a flat surface (of stainless steel, glass, a polymeric film or paper); a doctor blade spreads the slip into a thin tape of uniform thickness.

In the drying process, components are removed by evaporation, this green product is a flexible tape that may be cut or into which holes may be punched prior to a firing operation. Tape thicknesses normally range between 0.1 and 2 mm. Tape casting is widely used in the production of ceramic substrates that are used for integrated circuits and for multilayered capacitors.

10.5 Drying

A ceramic piece that has been formed hydroplastically or by slip casting retains significant porosity and insufficient strength for most practical applications. In addition, it may still contain some liquid (e. g., water), which was added to assist in the forming operation. This liquid is removed in a drying process.

As a clay-based ceramic body dries, it also experiences some shrinkage. In the early stages of drying the clay particles are virtually surrounded by and separated from one another by a thin film of water. As drying progresses and water is removed, the interparticle separation decreases, which is manifested as shrinkage. During drying it is critical to control the rate of water removal. Drying at interior regions of a body is accomplished by the diffusion of water molecules to the surface where evaporation occurs. If the rate of evaporation is greater than the rate of diffusion, the surface will dry (and as a consequence shrink) more rapidly than the interior, with a high probability of the formation of the defects.

10.6 Firing

After drying, a body is usually fired at an elevated temperature; the firing temperature depends on the composition and desired properties of the finished piece. During the firing operation, the density of the formed piece is further increased (with an attendant decrease in porosity) and the mechanical strength is enhanced. These changes occur by the coalescence of the powder particles into a more dense mass in a process termed sintering. The mechanism of sintering is schematically illustrated in Figure 10.4. After pressing, many of the powder particles touch one another (Figure 10.4a). During the initial sintering stage, necks form along the contact regions between adjacent particles, in addition, a grain boundary forms within each neck, and every interstice between particles becomes a pore (Figure 10.4b). As sintering progresses, the pores become smaller and more

Chapter 10 Ceramic Processing Methods

spherical in shape (Figure 10.4c).

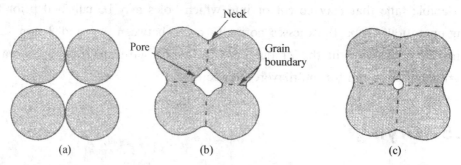

Figure 10.4 For a powder compact, microstructural changes that occur during firing. Powder particles after pressing (a), particle coalescence and pore formation as sintering begins (b) and as sintering proceeds, the pores change size and shape (c)

Processing of ceramics is carried out through the consolidation of loose powders to form polycrystalline objects. The majority of ceramics are sintered, using solid-state sintering or liquid-phase sintering. Generally, sintering occurs at temperatures below the melting point of single-phase ceramics or below their liquids for multiphase ceramics, although a liquid phase may be present temporarily during sintering. Therefore, mass transport necessary to effect the changes shown in Figure 10.4 is accomplished by atomic diffusion from the bulk particles to the neck regions.

Notes

① motor-driven auger 发动机传动的螺旋   ② crack-initiation 裂纹萌生；裂纹开裂

Vocabulary

auger [ˈɔːgə] n. [机]螺旋钻；[木]木螺钻；地螺钻 vt. 用钻子钻洞于
beneficiate [ˌbenɪˈfɪʃɪeɪt] vt. 为改善性能而进行的处理
ceramic [sɪˈræmɪk] n. 陶瓷；陶瓷制品 adj. 陶瓷的；陶器的；制陶艺术的
coalescence [ˌkəʊəˈlesns] n. 合并；联合；接合
consolidation [kənˌsɒlɪˈdeɪʃən] n. 巩固；合并；团结

deairing [diːˈaɪrɪŋ] n. 去气；真空除气（黏土材料）
flatness [ˈflætnɪs] n. 平坦；单调；断然的态度
grindiny [ˈgraɪndɪŋ] adj. 磨的；刺耳的；令人难以忍受的
impractical [ɪmˈpræktɪk(ə)l] adj. 不切实际的,不现实的；不能实行的
ingredient [ɪnˈgriːdɪənt] n. 原料；要素；组成部分 adj. 构成组成部分的

Chapter 10　Ceramic Processing Methods

interparticle [ˌɪntɜːrpɑːtɪkl] adj. 粒子间的；颗粒间的
isostatic [ˌaɪsəʊˈstætɪk] adj. 均衡说的；[地物]地壳均衡的
lubricate [ˈluːbrɪkeɪt] vt. 使……润滑；给……加润滑油　vi. 润滑；涂油；起润滑剂作用
particulate [pɑːˈtɪkjʊlət] n. 微粒，微粒状物质　adj. 微粒的
paste [peɪst] n. 面团，膏；糊状物，[胶粘]浆糊　vt. 张贴；裱糊；用浆糊粘
pore [pɔː] n. 气孔，小孔
porosity [pɔːˈrɒsɪti] n. 有孔性，多孔性

refractory [rɪˈfrækt(ə)rɪ] n. 倔强的人；耐火物质　adj. 难治的；难熔的；不听话的
shrinkage [ˈʃrɪŋkɪdʒ] n. 收缩；减低
suspension [səˈspenʃ(ə)n] n. 悬浮；暂停；停职
taxonomical [ˌtæksəʊˈnɒmɪkəl] adj. 分类学的
uniaxial [ˌjuːnɪˈæksɪəl] adj. 单轴的
viscosity [vɪˈskɒsɪti] n. [物]粘性，[物]粘度
viscous [ˈvɪskəs] adj. 粘性的；黏的

Exercises

1. Translate the following Chinese phrases into English

(1) 陶瓷成型技术　　(2) 塑料陶瓷　　(3) 原材料
(4) 空心注浆　　(5) 不切实际的温度　　(6) 干燥过程
(7) 液相烧结

2. Translate the following English phrases into Chinese

(1) hot rolling　　　　(2) nonplastic materials
(3) solid casting　　　(4) hot pressing
(5) tape casting　　　(6) initial sintering stage

3. Translate the following Chinese sentences into English

(1) 因为它们是作为结构件使用的，经常承受机械载荷，所以被称为结构陶瓷。
(2) 陶瓷的多段制造过程包括粉末生产、粉末调节、成型、干燥和致密化(也叫做焙烧或烧结)。
(3) 陶瓷注射成形是先进陶瓷的近净成形技术。
(4) 我们首先需要说明的是"什么是玻璃"。传统的观点认为玻璃是从熔融的液体淬冷得到的固体。
(5) 陶瓷元件是将仔细研磨的材料经过加压和烧结制成的。

4. Translate the following English sentences into Chinese

(1) You may find that, in addition to dividing ceramics according to their properties and applications, it is common to class them as traditional or advanced.

(2) Since, in ancient times, the potter was significantly associated with clay work, as such ceramics meant art of potter dealing with clay and clay article fired to give hardness.

Chapter 10 Ceramic Processing Methods

(3) Since ceramics are often oxides, nitrides, or sulfides, compounds found in nature, there is little driving force for corrosion.

(4) The electrical nature of glass which lacks electronic conductivity is that it can support electric potential without free electron movement.

5. Translate the following Chinese essay into English

按照年产值增加的顺序排列,陶瓷工业所包括的主要方面有:(1) 研磨剂;(2) 搪瓷涂层;(3) 耐火材料;(4) 卫生瓷;(5) 结构性黏土产品;(6) 电子陶瓷和技术陶瓷产品;(7) 玻璃。

6. Translate the following English essay into Chinese

Clay is essentially a hydrated compound of aluminum and silicon $H_2Al_2Si_2O_9$, containing more or less foreign matter such as ferric oxide Fe_2O_3, which contributes the reddish color frequently associated with clay; silica SiO_2 as sand; and calcium carbonate $CaCO_3$ as limestone, *etc*.

扫一扫,查看更多资料

Chapter 11 Polymer Synthesis

11.1 Introduction

A polymer is a material made up of small repeating structural units combined to give a very large, linear structure. These ultra-large molecules are called macromolecules. The small repeating units found within polymers come from the individual starting molecules, which are called monomers. All commercially important polymers are made up of hundreds or thousands of monomer units joined together to form one large molecule. Long before the first synthetic polymer was invented, we used natural polymers, such as cotton, silk and wool, for our clothing. Like synthetic polymers, these nature polymers consist of many of the same small molecules joined together in a long chain to make macromolecules. Regardless of their ultimate source, what all polymers have in common is that they are made up of very large, linear molecules containing repeating groups within the chain. The size of the individual polymer molecules and the nature of the groups ultimately controls what physical properties the bulk material will have. In many cases (e. g., polyethylene, polyvinyl chloride and polystyrene) the monomer molecules contain C = C double bonds that are broken during the polymerization process. These are referred to as "chain-growth polymers". In other cases, the monomer molecules may connect while forming a molecule of water (e. g., nylon or polyethylene terephthalate). These are called "step-growth polymers". Generic examples of each type of process are shown below (Figure 11.1).

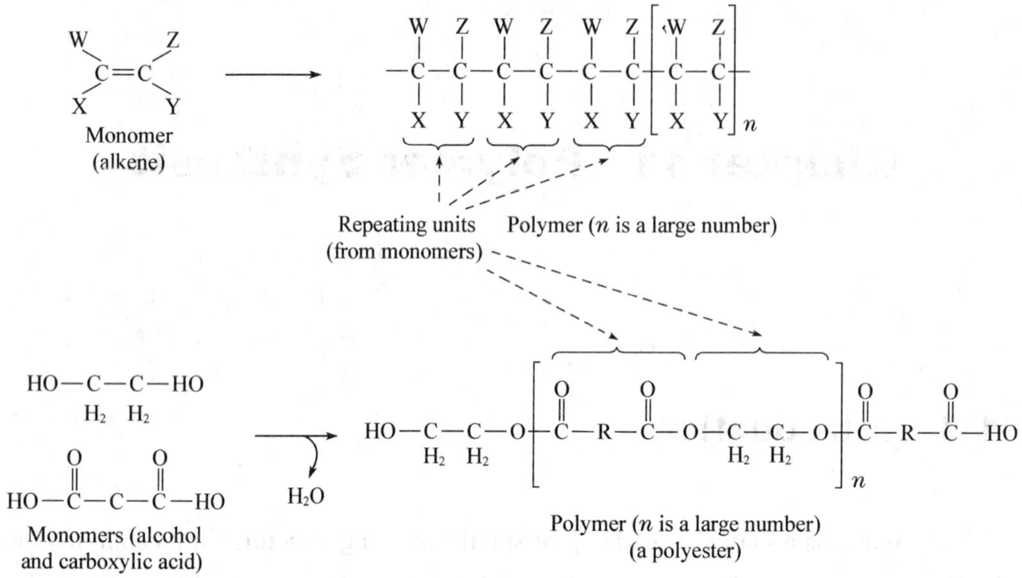

Chapter 11 Polymer Synthesis

Figure 11.1 Generic examples of polymer formation from gonomers

11.2 Polyethylene Synthesis

The simplest synthetic polymer is polyethylene. Recalling that ethylene is the old name for ethene, the simplest example of an alkene (Figure 11.2). Polyethylene is a chain-growth polymer, in which the double bond of the alkene monomer gets broken as new bonds are formed between monomer molecules. Can you predict the basic structure of polyethylene? As you might imagine, it simply consists of -CH_2CH_2- units repeated again and again. The most common type of chain-growth polymerization uses "radical initiation" in which a small amount of a compound with only seven valence electrons (R., called a free radical) begins the polymerization by adding to an alkene monomer and generating a new radical that then adds to another alkene, and so on. The polymer is "terminated" by reaction with another radical, either another initiator or another growing chain.

Polyethylene polymerization mechanism is as follows:

(1) Initiation

A free radical is generated from the thermal decomposition of a peroxide or by an as-yet unknown mechanism in the case of oxygen. The free radical then reacts with ethylene to start a growing polymer chain.

$$I\cdot + CH_2=CH_2 \longrightarrow I-CH_2-CH_2\cdot$$

$$R_i = k_i[I]$$

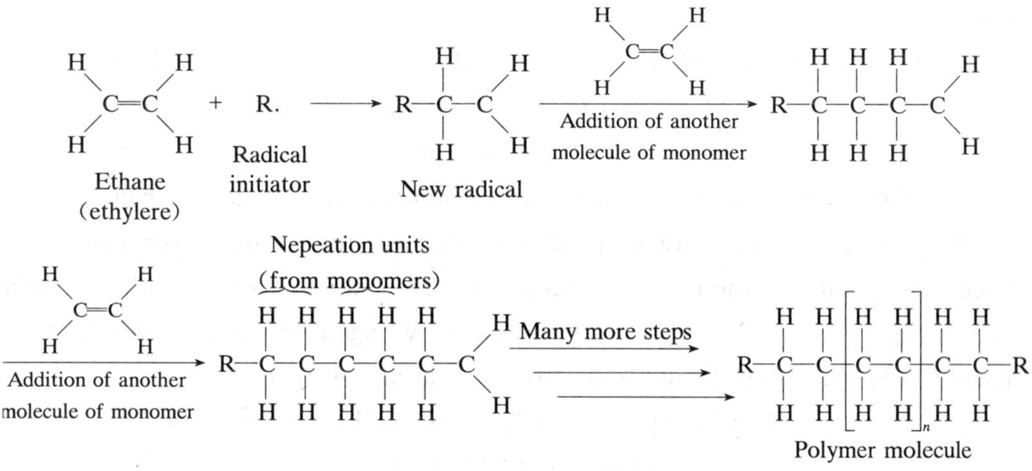

Figure 11.2 Polymerization of ethylene to form polyethylene

The choice of initiator is made on the basis of the type of reactor, the residence time in each zone, the desired reaction temperature, and initiator cost. A principal factor in choosing an initiator is the half-life[①]. The longer the half-life of the initiator, the higher the degree of polymerization, but the higher the cost of the peroxide.

(2) Chain propagation

Chain propagation proceeds by radical reaction with ethylene or comonomer molecules. In the case of ethylene homopolymerization, the mechanism and kinetics are straightforward.

$$R-CH_2-CH_2 \cdot + CH_2=CH_2 \longrightarrow R-CH_2-CH_2-CH_2 \cdot$$
$$R_p = k_p[R \cdot]P_e \text{ or } R_p = k_p[R \cdot][E]$$

where P_e is the ethylene pressure and E is ethylene.

(3) Termination

There are a variety of competing mechanisms by which the growth of a polymer chain can be stopped. The dominant mechanisms are determined by the polymerization conditions and the concentrations of chain-transfer agents present.

(4) Termination by coupling

Two growing polymer chains can react together to form one long polymer molecule or can disproportionate to form two inert chains.

$$R-CH_2-CH_2 \cdot + R-CH_2 \cdot \longrightarrow R-CH_2-CH_2-CH_2-R$$
$$\text{or } R-CH_3 + R-CH=CH_2 \quad R_t = k_t[R \cdot]^2$$

(5) Termination by chain transfer with ethylene

The growing polymer chain can react with ethylene, and instead of the ethylene inserting into the growing chain, a radical transfer reaction can take place to form vinyl end group.

$$R-CH_2-CH_2 \cdot + CH_2 = CH_2 \longrightarrow RCH_2 = CH_2 + CH_2-CH_2 \cdot$$
$$\text{or } RCH_2CH_3 + CH_2 = CH \cdot$$
$$R_{tre} = k_{tre}[R \cdot]P_e$$

(6) Termination by chain transfer with chain-transfer agents or solvents

Polyethylene radicals are very reactive and will react with solvents or other trace contaminants in the reactor to terminate one chain and begin another. Chain transfer agents, such as propane, propylene, hydrogen and isobutylene, can be added to the reaction to facilitate control of molecular weight.

$$R-CH_2-CH_2 \cdot + SH \longrightarrow RCH_2CH_3 + SH \cdot$$
$$\text{or } RCH_2CH_3 + CH_2 = CH \cdot$$
$$R_{trs} = k_{trs}[R \cdot][S]$$

Chain-transfer constants for some common chain-transfer agents, as well as impurities found in feedstocks, are found in Table 11.1. As resin producers are striving to increase production rates and aim-grade polymer yield, reaction models are being used to predict polymer properties and accelerate transitions. Analytical devices are being used to measure the impurities in the incoming and recycled ethylene streams and the corresponding chain-transfer constants applied to calculate. The molecular weights are made in the reactor. Therefore, these chain-transfer constants are more than just of academic interest.

Table 11.1 Common chain-transfer agents and their constants

Agent	Transfer	Reference
Methane	0.000 2	15
Ethane	0.006 0	15
Propane	0.003 0	16
Ethanol	0.007 5	17
Propylene	0.012 2	17
Hydrogen	0.015 9	17
Acetone	0.015 8	17

11.3 Nylon Synthesis

Nylon was one of the first big success stories in the field of polymer chemistry. Nylon is an example of a polyamide. You know what polyester looks like from the discussion above; if you refer to your table of functional groups, can you predict what a polyamide should look like? Actually, the most important polyamides are the proteins, and it was from proteins that Wallace Carrothers, a chemist at Dupont, took his inspiration. He found that mixing two monomers, adipic acid (a molecule with two carboxylic acids as we saw with terephthalic acid) and hexamethylene diamine (a molecule with two amine groups, similar to ethylene glycol), gave the desired polymer (Figure 11.3). As with polyesters, a molecule of water is generated each time a monomer adds to the growing chain. Nylon's importance quickly became apparent when it was discovered how easily it could be spun into strong, smooth fibers. It was first used in the bristles of toothbrushes. The resemblance of this material to silk led to the notion of making stockings from Nylon, and on the first day that Nylon stockings were offered for sale in New York City in 1940, four million pairs were bought. Unfortunately for the avid consumers, the supply of Nylon quickly dried up after the United States entered world War Ⅱ. It found many uses, including clothing, ropes and parachutes. It is still used for material that requires great strength and durability.

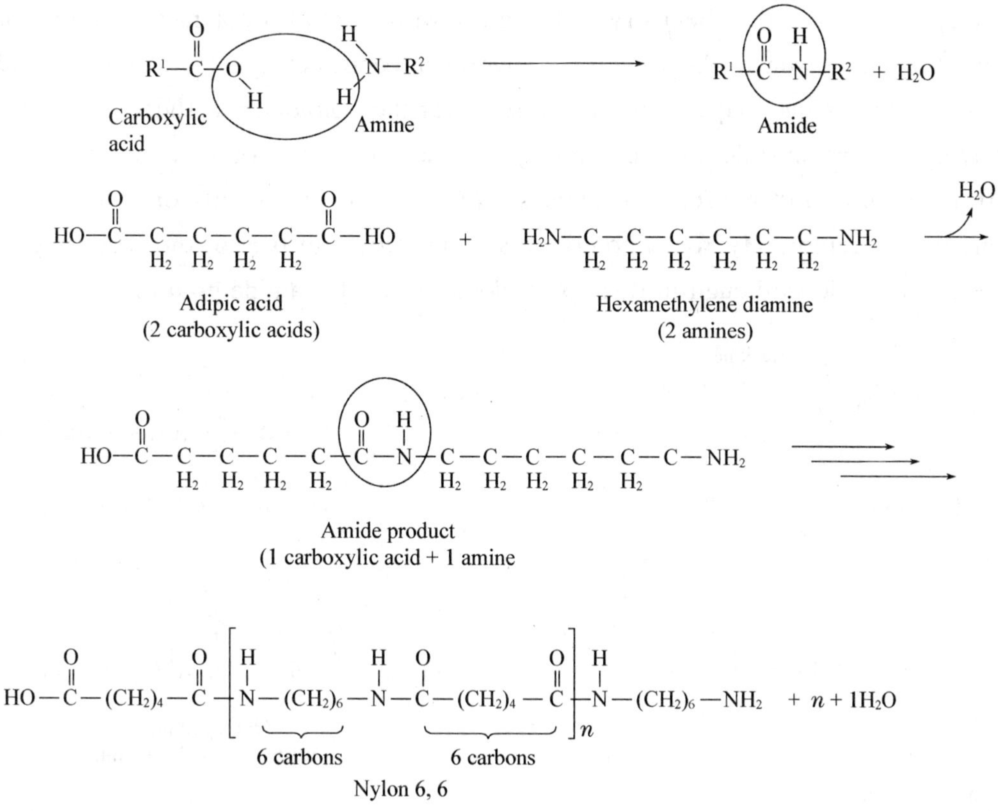

Figure 11.3 Nylon-6, 6 from condensation polymerization of adipic acid and hexamethylene diamine

The particular type of nylon invented by Carrothers and Dupont was called nylon-6-6 due to the number of carbons in each of the monomer pieces, and was such a commercial hit that other chemical companies tried to develop alternative products that did not infringe on the Dupont patent. A way around this was found with the closely related polymer, nylon-6 (Figure 11.4). Unlike nylon-6-6, nylon-6 is made from a single monomer that contains an amide functional group within a ring. When this amide, reacts with water and acid (H^+), it is converted into a molecule with one carboxylic acid and one amine. This molecule can then react with more of the caprolactam to give a new amide that has a carboxylic acid and an amine at the two ends, and eventually a polymer that looks very much like nylon-6-6. However, since there is only one monomer, having six carbon atoms, it is called nylon-6. One of the main reasons for the high tensile strength of the nylons (the property that allows very strong fibers to be made from them) is the ability of the individual polymer chains to form connections between each other (Figure 11.5). Amides such as those found in the nylons can form what is called hydrogen bonds. These connections are weaker than a full fledged covalent bond, but formation of several of them between two chains will hold them very tightly together. The same principle is seen in the chemistry of proteins, which are also polymers made up of repeating amide bonds. As we will see, hydrogen bonds between amide groups hold proteins into unique shapes that are necessary for their biological activity. They also play an important role in the binding together of two protein molecules, the attraction of a substrate for an enzyme's active site, or the affinity of a drug for a particular receptor. Hydrogen bonds are also an important part of the chemistry of the genetic code, although in this case it does not involve amide groups.

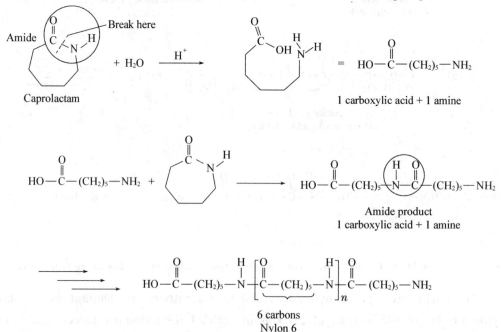

Figure 11.4 Formation of Nylon-6 from caprolactam

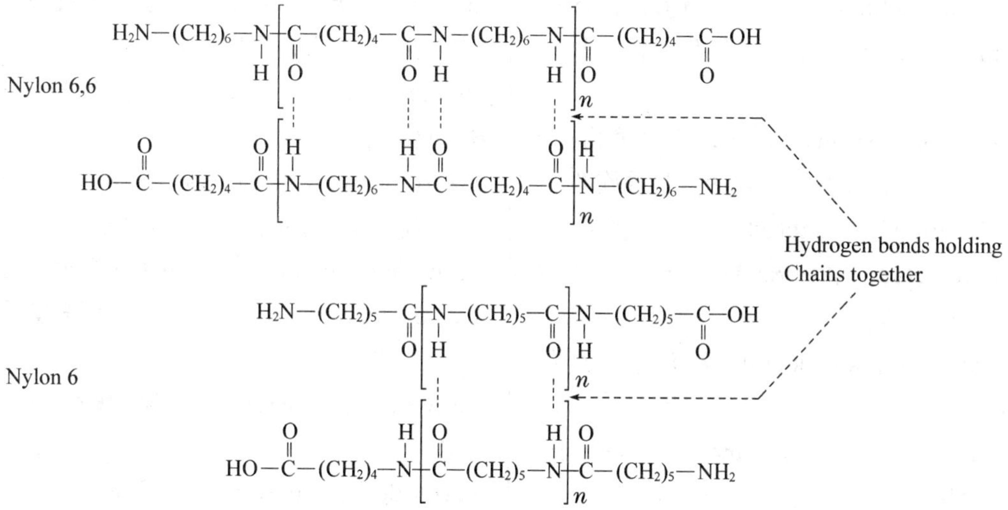

Figure 11.5 Hydrogen bonds between chains of Nylon-6, 6 or Nylon-6

One final point concerning polyamides such as the nylons concerns a structurally similar polymer, Kevlar. Kevlar is an amazingly strong material when spun into fibers; in fact, its tensile strength is greater than that of steel. Yet, it is much lighter than steel or other metals known for their strength. As a result, it is used in bulletproof vests, army helmets and protective clothing worn by firefighters. One of the monomers used to make Kevlar should be familiar to you, terephthalic acid (remember PET). The other is a compound with two amine groups, like the compound hexamethylene diamine used in nylon-6-6. However, in this case the two amine groups are attached directly to an aromatic ring. The two monomers come together in the same way as we saw before with nylon-6-6, forming a long chain in which the individual monomers are connected by amide linkages. The aromatic rings in both pieces make the polymer molecule very rigid, but the individual polymer chains can still bind to each other with hydrogen bonds, as we saw with the nylons. The result is a very stiff, strong material. In fact, the biggest challenge with Kevlar is how to fabricate it and it does not melt until it is heated above 500 ℃ (912 ℉).

Notes

① half-life 半衰期

Vocabulary

adipic [əˈdɪpɪk] *adj.* 脂肪的,油质的
acetone [ˈæsɪtəʊn] *n.* [有化]丙酮

affinity [əˈfɪnɪtɪ] *n.* 密切关系;吸引力;姻亲关系;类同

Chapter 11 Polymer Synthesis

alkene ['ælkiːn] n. [有化]烯烃;链烯烃
amine [ə'miːn] n. [有化]胺（等于 amin）
aromatic [ærə'mætɪk] n. 芳香植物;芳香剂 adj. 芳香的,芬芳的;芳香族的
bulletproof ['bʊlɪtpruːf] adj. 防弹的
caprolactam [ˌkæprəʊ'læktəm] n. [有化]己内酰胺
carboxylic [ˌkɑːbɒk'sɪlɪk] adj. [有化]羧基的
comonomer [kəʊ'mɒnəmə] n. [高分子]共聚用单体
contaminant [kən'tæmɪnənt] n. 污染物;致污物
diamine [daɪ'eɪmiːn] n. 二元胺;[无化]联氨
disproportionate [ˌdɪsprə'pɔːʃ(ə)nət] adj. 不成比例的
durability [ˌdjʊərə'bɪlɪtɪ] n. 耐久性;坚固;耐用年限
enzyme ['enzaɪm] n. [生化]酶
ethane ['iːθeɪn] n. [有化]乙烷
ethylene ['eθɪliːn] n. 乙烯
feedstock ['fiːdstɒk] n. 原料;给料（指供送入机器或加工厂的原料）
generic [dʒɪ'nerɪk] adj. 类的;一般的;属的;非商标的
glycol ['ɡlaɪkɒl] n. 乙二醇;甘醇;二羟基醇
homopolymerization [ˌhəʊmɒpɒlɪməraɪ'zeɪʃən] n. 均聚;同聚反应
infringe [ɪn'frɪn(d)ʒ] vt. 侵犯;违反;破坏 vi. 侵犯;侵害

initiator [ɪ'nɪʃɪeɪtə] n. 发起人,创始者;教导者;[计]启动程序;引爆器
linkage ['lɪnkɪdʒ] n. 连接;结合;联接;联动装置
macromolecule [ˌmækrə(ʊ)'mɒlɪkjuːl] n. [高分子]高分子;[化学]大分子
monomer ['mɒnəmə] n. 单体;单元结构
nylon ['naɪlɒn] n. 尼龙,[纺]聚酰胺纤维;尼龙袜
parachute ['pærəʃuːt] n. 降落伞 vi. 跳伞
peroxide [pə'rɒksaɪd] n. 过氧化氢;过氧化物 vt. 以过氧化氢漂白;以过氧化物处理 adj. 以过氧化氢漂白的
polyamide [ˌpɒlɪ'eɪmaɪd] n. [高分子]聚酰胺（尼龙）
polymer ['pɒlɪmə] n. [高分子]聚合物
polymerization [ˌpɒlɪməraɪ'zeɪʃn] n. 聚合;[高分子]聚合作用
propagation [ˌprɒpə'ɡeɪʃən] n. 传播;繁殖;增殖
receptor [rɪ'septə] [生化]受体;接受器;感觉器官
resemblance [rɪ'zembl(ə)ns] n. 相似;相似之处;相似物
resin ['rezɪn] n. 树脂;松香 vt. 涂树脂;用树脂处理
terephthalate [ˌterəf'θæleɪt] n. 对苯二酸盐;对苯二酸酯
termination [tɜːmɪ'neɪʃ(ə)n] n. 结束,终止
vinyl ['vaɪnɪl] n. 乙烯基

Exercises

1. Translate the following Chinese phrases into English

(1) 聚合物合成 (2) 天然聚合物 (3) 生长链
(4) 链增长 (5) 抗拉强度 (6) 酰胺键

(7) 单个高分子链

2. Translate the following English phrases into Chinese

(1) large molecule (2) polymerization process

(3) polyethylene synthesis (4) chain-transfer agent

(5) hydrogen bond (6) aromatic ring

3. Translate the following Chinese sentences into English

(1) 人们通常研究聚合物的领域是聚合物化学、聚合物物理和聚合物科学。

(2) 聚合物包括天然材料,例如橡胶和合成材料。

(3) 在聚合物中,交叉连接的数目越多,材料的刚性越强。

(4) 聚乙烯是结构最简单的高分子,也是应用最广泛的高分子材料。

4. Translate the following English sentences into Chinese

(1) Like any molecule, a polymer molecule's size may be described in terms of molecular weight or mass.

(2) During the past 70 years or so, chemists have learned to form synthetic polymers by polymerizing monomers through controlled chemical reactions.

(3) A single polymer molecule may consist of hundreds to a million monomers and may have a linear, branched or network structure.

(4) Two samples of natural rubber may exhibit different durability even though their molecules comprise the same monomers.

(5) Chain stiffness is also greatly increased when a ring is incorporated in the chain, as this restricts the rotation in the backbone and reduces the number of conformations a polymer can adopt.

5. Translate the following Chinese essay into English

热塑性材料通常为粒料,加工时将其置入经电加热并保持恒温的机筒中。采用这种简单的设备,机筒壁传导热量将材料融化。由于塑料热导性较差,传热过程需要一定的时间,因此塑料在机筒壁处可能会降解。

6. Translate the following English essay into Chinese

Polymeric nanocomposites can be considered as an important category of organicinorganic hybrid materials, in which inorganic nanoscale building blocks (e.g., nanoparticles, nanotubes, or nanometer thick sheets) are dispersed in an organic polymer matrix. They represent the current trend in developing novel nanostructured materials. When compared to conventional composites based on micrometer-sized fillers, the interface between the filler particles and the matrix in a polymer nanocomposite constitutes a much greater area within the bulk material, and hence influences the composite's properties to a much greater extent, even at a rather low filler loading.

扫一扫,查看更多资料

Chapter 12　Metal Matrix Composite

12.1　Introduction

Metal matrix composites (MMCs), like all composites, consist of at least two chemically and physically distinct phases, suitably distributed to provide properties not obtainable with either of the individual phases. Generally, there are two phases, e. g., a fibrous or particulate phase, distributed in a metallic matrix. Examples include continuous Al_2O_3 fiber reinforced Al matrix composites used in power transmission lines, Nb-Ti filaments in a copper matrix for superconducting magnets; tungsten carbide (WC)/cobalt (Co) particulate composites used as cutting tool and oil drilling inserts, and SiC particle reinforced Al matrix composites used in aerospace, automotive and thermal management applications.

A legitimate question that the reader might ask is: Why metal matrix composites? The answer to this question can be subdivided into two parts: advantages with respect to unreinforced metals and advantages with respect to other composites such as polymer matrix composites (PMCs). With respect to metals, MMCs offer the following advantages:

① Major weight savings due to higher strength-to-weight ratio.
② Exceptional dimensional stability (compare, for example, SiC/Al to Al).
③ Higher elevated temperature stability, i. e., creep resistance.
④ Significantly improved cyclic fatigue characteristics.

With respect to PMCs, MMCs offer these distinct advantages:

① Higher strength and stiffness.
② Higher service temperatures.
③ Higher electrical conductivity (grounding, space charging).
④ Higher thermal conductivity.
⑤ Better transverse properties.
⑥ Improved joining characteristics.

Chapter 12 Metal Matrix Composite

⑦ Radiation survivability (laser, UV, nuclear, *etc.*).

⑧ Little or no contamination (no out-gassing① or moisture absorption problems).

(1) Types of MMCs

All metal matrix composites have a metal or a metallic alloy as the matrix. The reinforcement can be metallic or ceramic. In some unusual cases, the composite may consist of a metallic alloy "reinforced" by a 6 ber reinforced polymer matrix composite (*e. g.*, a sheet of glass fiber reinforced epoxy or aramid fiber reinforced epoxy).

In general, there are three kinds of metal matrix composites (MMCs): particle reinforced MMCs; short fiber or whisker reinforced MMCs; continuous fiber or sheet reinforced MMCs.

Figure 12.1 shows, schematically, the three major types of metal matrix composites: Continuous fiber reinforced, short fiber or whisker reinforced, particle reinforced, and laminated or layered composites. The reader can easily visualize that the continuous fiber reinforced composites will be the most anisotropic of all. Table 12.1 provides examples of some important reinforcements used in metal matrix composites as well as their aspect ratios (length/diameter) and diameters.

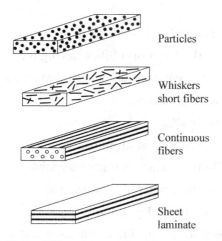

Figure 12.1 Different types of metal matrix composites

Table 12.1 Typical reinforcements used in metal matrix composites

Type	Aspect ratio	Diameter/μm	Examples
Particle	1~4	1~25	SiC, Al_2O_3, BN, B_4C, WC
Short fiber or whisker	10~10 000	1~5	C, SiC, Al_2O_3, $Al_2O_3 + SiO_2$
Continuous fiber	>1 000	3~150	Si, Al_2O_3, C, B, W, Nb-Ti, Nb_3Sn

Chapter 12 Metal Matrix Composite

Particle or discontinuously reinforced MMCs (the term discontinuously reinforced MMCs is commonly used to indicate metal matrix composites having reinforcements in the form of short fibers, whiskers or particles) have assumed special importance for the following reasons:

① Particle reinforced composites are inexpensive vis-a-vis continuous fiber reinforced composites. Cost is an important and essential parameter, particularly in applications where large volumes are required (e. g., automotive applications).

② Conventional metallurgical processing techniques such as casting or powder metallurgy, followed by conventional secondary processing by rolling, forging, and extrusion can be used.

③ Higher use temperatures than the unreinforced metal.

④ Enhanced modulus and strength.

⑤ Increased thermal stability.

⑥ Better wear resistance.

⑦ Relatively isotropic properties compared to fiber reinforced composites.

Within the broad category of discontinuously reinforced composites, metal matrix composites made by liquid metal casting are somewhat cheaper to produce than powder metallurgy composites. There are two types of cast metal matrix composites.

① Cast composites having local reinforcement.

② Cast composites in the form of a billet having uniform reinforcement with a wrought alloy matrix. Such composite billets are forged and/or extruded, followed by rolling or other forming operations.

(2) Characteristics of MMCs

One of the driving forces for metal matrix composites is, of course, enhanced stiffness and strength. There is other characteristics which may be equally valuable. As examples, we can cite the ability to control thermal expansion in applications involving electronic packaging. By adding ceramic reinforcements, one can generally reduce the coefficient of linear thermal expansion of the composite. Electrical and thermal conductivity characteristics may be important in some applications. Clearly, superconductors require superconducting characteristics. The metallic matrix provides a high thermal conductivity medium in case of an accidental quench, in addition to holding the tiny superconducting filaments together. Other important characteristics that may be of immense value include wear resistance (e. g., in WC/Co composites used in cutting tools or oil drilling inserts and SiCdAl rotor in brakes). Thus, although one commonly uses the term reinforcement by particle or fibers in the context of metal matrix composites, it is worth pointing out that strength enhancement may not be the most important characteristic in many applications. In the chapters that follow we explore these and other unique and important attributes of metal matrix composites.

(3) Composition of MMCs

MMCs are made by dispersing a reinforcing material into a metal matrix. The reinforcement surface can be coated to prevent a chemical reaction with the matrix. For example, carbon fibers are commonly used in aluminium matrix to synthesize composites showing low density and high strength. However, carbon reacts with aluminium to generate a brittle and water-soluble[②] compound Al_4C_3 on the surface of the fiber. To prevent this reaction, the carbon fibers are coated with nickel or titanium boride.

① Matrix

The matrix is the monolithic material into which the reinforcement is embedded, and is completely continuous. This means that there is a path through the matrix to any point in the material, unlike two materials sandwiched together. In structural applications, the matrix is usually a lighter metal such as aluminium, magnesium or titanium, and provides a compliant support for the reinforcement. In high temperature applications, cobalt and cobalt-nickel alloy matrices are common.

② Reinforcement

The reinforcement material is embedded into the matrix. The reinforcement does not always serve a purely structural task (reinforcing the compound), but is also used to change physical properties such as wear resistance, friction coefficient or thermal conductivity. The reinforcement can be either continuous or discontinuous. Discontinuous MMCs can be isotropic, and can be worked with standard metalworking techniques, such as extrusion, forging or rolling. In addition, they may be machined using conventional techniques, but commonly would need the use of polycrystalline diamond tooling.

Continuous reinforcement uses monofilament wires or fibers such as carbon fiber or silicon carbide. Because the fibers are embedded into the matrix in a certain direction, the result is an anisotropic structure in which the alignment of the material affects its strength. One of the first MMCs used boron filament as reinforcement. Discontinuous reinforcement uses "whiskers", short fibers or particles. The most common reinforcing materials in this category are alumina and silicon carbide.

(4) Manufacturing and forming methods of MMCs

MMCs manufacturing can be broken into three types: solid, liquid and vapor.

① Solid state methods

(a) Powder blending and consolidation (powder metallurgy): Powdered metal and discontinuous reinforcement are mixed and then bonded through a process of compaction, degassing and thermo-mechanical treatment (possibly via hot isostatic pressing (HIP) or extrusion).

(b) Foil diffusion bonding: Layers of metal foil are sandwiched with long fibers, and then pressed through to form a matrix.

② Liquid state methods

(a) Electroplating/electroforming

A solution containing metal ions loaded with reinforcing particles is co-deposited forming a composite material.

(b) Stir casting

Discontinuous reinforcement is stirred into molten metal, which is allowed to solidify.

(c) Squeeze casting

Molten metal is injected into a form with fibers preplaced inside it.

(d) Spray deposition

Molten metal is sprayed onto a continuous fiber substrate.

(e) Reactive processing

A chemical reaction occurs, with one of the reactants forming the matrix and the other the reinforcement.

③ Vapor deposition

(a) Physical vapor deposition

The fiber is passed through a thick cloud of vaporized metal, coating it.

(b) In situ fabrication technique

(c) Controlled unidirectional solidification of a eutectic alloy can result in a two-phase microstructure with one of the phases, present in lamellar or fiber form, distributed in the matrix.

12.2 Processing of Metal Matrix Composite

Metal matrix composites can be made by liquid, solid or gaseous state processes. In this section we describe some important processing techniques for fabricating MMCs.

(1) Liquid state processing

Metal matrix composites can be processed by incorporating or combining a liquid metal matrix with the reinforcement. There are several advantages to using a liquid phase route in processing. These include near net-shape (when compared to solid state processes like extrusion or diffusion bonding), faster rate of processing, and the relatively low temperatures associated with melting most light metals, such as Al and Mg. The most common liquid phase processing techniques can be subdivided into four major categories.

① Casting or liquid infiltration

This involves infiltration of a fibrous or particulate preform by a liquid metal. In the case of direct introduction of short fibers or particles into a liquid mixture, consisting of liquid metal and ceramic particles or short fibers, is often stirred to obtain a homogeneous distribution of particles. In centrifugal casting, a gradient in reinforcement particle loading is obtained. This can be quite advantageous from a machining; or performance perspective.

② Squeeze casting or pressure infiltration

This method encompasses pressure-assisted liquid infiltration of a fibrous or particulate preform. This process is particularly suited for complex shaped components, selective or localized reinforcement, and where production speed is critical.

(2) Solid state processing

The main drawback associated with liquid phase techniques is the difficulty in controlling reinforcement distribution and obtaining a uniform matrix microstructure. Furthermore, adverse interfacial reactions between the matrix and the reinforcement are likely to occur at the high temperatures involved in liquid processing. These reactions can have an adverse effect on the mechanical properties of the composite. The most common solid phase processes are based on powder metallurgy techniques. These typically involve discontinuous reinforcements, due to the ease of mixing and blending, and the effectiveness of densification. The ceramic and metal powders are mixed, isostatically cold compacted, and hot pressed to full density. The fully-dense compact then typically undergoes a secondary operation such as extrusion or forging. Novel low-cost approaches, such as sinter-forging, have aimed at eliminating the hot pressing step with promising results.

① Powder metallurgy processing

Powder processing involves cold pressing and sintering, or hot pressing to fabricate primarily particle- or whisker-reinforced MMCs. The matrix and the reinforcement powders are blended to produce a homogeneous distribution. The blending stage is followed by cold pressing to produce what is called a green body[3], which is about 80% dense and can be easily handled, Figure 12.2. The cold pressed green body is canned in a container, sealed and degassed to remove any absorbed

Chapter 12　Metal Matrix Composite

moisture from the particle surfaces. One of the problems with bonding metallic powder particles, such as Al particles, to ceramic particles, such as Al particles, to ceramic particles, such as SiC, or to other Al particles are the oxide "skin" that is invariably present on the Al particle surface. Degassing and hot pressing in an inert atmosphere contributes to the removal of Al hydrides present on the particle surface, making the oxide skin more brittle and, thus, more easily sheared. The material is hot pressed, uniaxially or isostatically, to produce a fully dense composite and extruded. The rigid particles or fibers do not deform, causing the matrix to be deformed significantly.

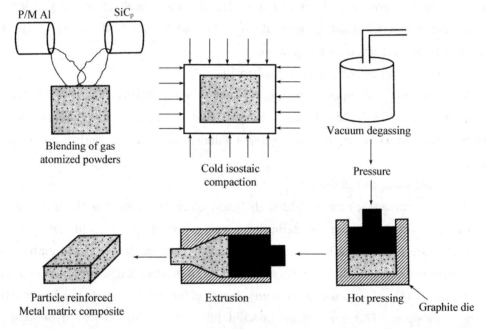

Figure 12.2　Powder processing, hotpressing and extrusion process for fabricating particulate or short fiber reinforced MMCs

② Extrusion

Extrusion processing has been used extensively as a means of secondary deformation processing of MMCs. It is particularly advantageous because the combination of pressure and temperature results in shear between Al/Al particles and Al/SiC particles, which contributes to fracture of the oxide skin on the Al particles, and the bonding between particle and matrix is enhanced. Because of the large strains associated with this process, however, extrusion has been used primarily to consolidate composites with discontinuous reinforcement, in order to minimize reinforcement fracture. Even in discontinuously reinforced materials, fracture of short fibers or particles often takes place, which can be detrimental to the properties of the composite.

③ Forging

Forging is another common secondary deformation processing technique used to manufacture metal matrix composites. Once again, this technique is largely restricted to composites with discontinuous reinforcement. In conventional forging, a hot-pressed or extruded product is forged to near-net shape.

④ Pressing and sintering

A relatively inexpensive and simple technique involves pressing and sintering of powders. These composite systems are typically sintered in a temperature range to obtain some degree of liquid phase. The liquid phase flows through the pores in the compact resulting in densification of the composite (unless interfacial reaction takes place). Special mention should be made of WC/Co composites, commonly known as cemented carbides. They are really nothing but very high volume fraction of WC particles distributed in a soft cobalt matrix. These composites are used extensively in machining and rock and oil drilling operations.

⑤ Roll bonding and Co-extrusion

Roll bonding is a common technique used to produce a laminated composite consisting of different metals in layered form. Such composites are called sheet laminated metal-matrix composites. Roll bonding and hot pressing have also been used to make laminates of Al sheets and discontinuously reinforced MMCs. Figure 12.3 shows the roll bonding process of making a laminated MMC.

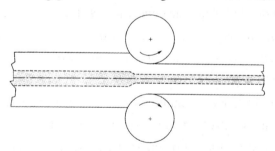

Figure 12.3 Roll bonding process of making a laminated MMC where a metallurgical bond is produced between the layers

⑥ Diffision bonding

Diffusion bonding is a common solid-state processing technique for joining; similar or dissimilar metals. Inter diffusion of atoms, at an elevated temperature, from clean metal surfaces in contact with each other leads to bonding. The principal advantages of this technique are the ability to process a wide variety of matrix metals and control of fiber orientation and volume fraction. Among the disadvantages are long processing times, high processing temperatures and pressures (which makes the process expensive), and limitation on complexity of shapes that can be produced.

⑦ Explosive shock consolidation

A fairly novel, high strain rate and rapid solidification technique is explosive shock consolidation. In this technique, dynamic compaction and/or synthesis of powders can be achieved by means of shock waves generated by either explosives in contact with the powder or high velocity impact from projectiles. This process is particularly attractive for consolidating hard materials such as ceramics or composites.

(3) Gaseous state processing

Plasma spraying is the primary form of gaseous state processing. The main application of plasma spraying was described above, to form matrix-coated fibers, which are subsequently hot-pressed to form the final product. In addition, laminated composites, particularly on the nanometer scale have been processed by physical vapor deposition (PVD) process. PVD processes (specifically, sputter deposition-based processes) offer an extremely wide range of possibilities for fabricating nanolaminate microstructures with tailored chemistry, structure, and thickness of the individual layers and interfaces. Additional important PVD processing parameters include reactive deposition, plasma-assisted deposition and substrate heating.

Some metal/metal layered systems, such as Ni/Cu and Ni/Ti, have been processed at the nanoscale by sputter deposition with great success. An example of a nanoscale Cu-Ni multilayer with a bilayer period of 5 nm was prepared. Note the well-defined[④] layered structure. The corresponding selected area diffraction pattern (inset) shows a $<001>$ growth direction and cube-oncube orientation relationship between FCC Cu and FCC Ni. A challenge in the synthesis of nanolaminates via the PVD approach is the control of intrinsic residual stresses. Some control over residual stresses has been achieved by energetic particle bombardment, either in situ or post-deposition[⑤] using an ion source. In the case of magnetron sputtering, a negative substrate bias may be sufficient to change the residual stress from tensile to compressive. A film deposited with low bombardment energy yields a tensile residual stress and a microstructure with nanoscale columnar porosity cracking for a sputtered 150 nm thick Cr film. The same material sputtered with a negative bias, on the other hand, yields a nanocrystalline film with an equiaxed grain structure, near-zero residual stress, and, thus, no intergranular porosity.

Chapter 12　Metal Matrix Composite

◆ Notes

① out-gassing 除气；脱气
② water-soluble 可溶于水的
③ green body 生坯；未烧试样
④ well-defined 定义明确的；界限清楚的
⑤ post-deposition 后沉积

◆ Vocabulary

aramid ['ærəmɪd] n. 芳香族聚酰酪；人造纤维之一种

bias ['baɪəs] n. 偏见；偏爱；斜纹；乖离率　adj. 偏斜的　vt. 使存偏见　adv. 偏斜地

bilayer ['baɪleɪə] n. 双分子层（膜）

bombardment [bɒm'bɑːdm(ə)nt] n. 轰炸；炮击

boride ['bɔːraɪd] n. [无化]硼化物

degass [diː'gæs] t. 给……排气（或除气，脱气），排除（或放出）……里的气体，排气，放气，消除（毒气等）的毒性；消除……的毒气

densification [ˌdensɪfɪ'keɪʃən] n. 密实化；封严；[化学]稠化

discontinuous [ˌdɪskən'tɪnjʊəs] adj. 不连续的；间断的

distributed [dɪ'strɪbjʊtɪd] adj. 分布式的，分散式的

epoxy [ɪ'pɒksɪ] n. 环氧基树脂　adj. 环氧的　vt. 用环氧树脂胶合

filament ['fɪləm(ə)nt] n. 灯丝；细丝；细线；单纤维

infiltration [ˌɪnfɪl'treɪʃən] n. 渗透；渗透物

interfacial [ˌɪntə'feɪʃ(ə)l] adj. 界面的

laminated ['læmɪneɪtɪd] adj. 层压的；层积的；薄板状的　v. 分成薄片；用薄片覆盖（laminate 的过去分词）

magnetron ['mægnɪtrɒn] n. [电子]磁控管

monolithic [ˌmɒnə'lɪθɪk] n. 单块集成电路，单片电路　adj. 整体的；巨石的，庞大的；完全统一的

multilayer ['mʌltɪleɪə] n. 多分子层　adj. 有多层的

particulate [pɑː'tɪkjʊlət] n. 微粒，微粒状物质　adj. 微粒的

projectile [prə(ʊ)'dʒektaɪl] adj. 抛射的；抛掷的；供抛射用的；（触角等）能伸出的　n. 射弹；抛射体；自动推进武器

sandwiched ['sændwɪdʒd] adj. 夹于两者之间的　v. 夹在中间（sandwich 的过去式和过去分词）

survivability [səˌvaɪvə'bɪlɪtɪ] n. 存活的能力，生命力

visualize ['vɪzjʊəlaɪz] vt. 形象，形象化；想像，设想　vi. 显现

whisker ['wɪskə] n. [晶体]晶须；胡须；腮须

◆ Exercises

1. Translate the following Chinese phrases into English

（1）金属基复合材料　　　　（2）抗蠕变性　　　　（3）金属合金

Chapter 12 Metal Matrix Composite

(4) 粉末冶金复合材料 (5) 耐磨性 (6) 热导率
(7) 固态法 (8) 喷射沉积 (9) 液态浸渗法
(10) 界面反应 (11) 均匀分布 (12) 残余应力
(13) 搅拌铸造

2. Translate the following English phrases into Chinese

(1) polymer matrix composites (2) cyclic fatigue characteristics
(3) fiber reinforced composites (4) wrought alloy
(5) friction coefficient (6) anisotropic structure
(7) diffusion bonding (8) physical vapor deposition
(9) squeeze casting (10) discontinuous reinforcement
(11) hot pressing (12) intergranular porosity

3. Translate the following Chinese sentences into English

(1) 据说,复合材料拥有两个相。

(2) 考虑产品成本效益的原因,采用接近最终成品尺寸的模型和加工重整过程,最小化结构单元的机械加工工艺。

(3) 粉末冶金工艺也使此前只能通过快凝粉末法才可行的基体合金成分和微结构精炼得以应用。

(4) 金属陶瓷是金属基复合材料的重要一类,这种陶瓷的难熔金属基体中分散着尺寸大于 $1\ \mu m$ 的陶瓷晶粒。

(5) 铁基和钢基体比较便宜而且能在高温下使用,前提是重量不是主要的考虑因素。

(6) 陶瓷-金属复合材料具备高硬度、高强度、高韧性以及低密度的优点,已被广泛应用于防弹领域。

4. Translate the following English sentences into Chinese

(1) In about the middle 1960s, a new group of composite materials, called advanced engineered composite materials (advanced composites), began to emerge.

(2) Careful attention must be paid to the dispersion of the reinforcement components, so that the reactivity of the components used is coordinated with the temperature of the melt and the duration of stirring, since reactions with the melt can lead to the dissolution of the reinforcement components.

(3) In gas pressure infiltration the response times are clearly longer than in squeeze casting, so that the materials must be carefully selected and coordinated, in order to be able to produce the appropriate composite material for the appropriate requirements.

(4) The material cost is the major problem that currently limits their uses, otherwise most of the metallic structural parts can be replaced with metal matrix composite parts to gain advantages.

(5) The bimetal composites were produced by combining the high chromium and midcarbon steel.

5. Translate the following Chinese essay into English

事实上,由于金属基复合材料的制备和工作条件的原因,沿着纤维/基体相界面出现了特殊物理过程,在界面区域能产生明显影响金属基复合材料力学性能的化合物和/或物相。

6. Translate the following English essay into Chinese

These materials may be utilized at higher service temperatures than their base metal counterparts; furthermore, the reinforcement may improve specific stiffness, specific strength, abrasion resistance, creep resistance, thermal conductivity and dimensional stability.

扫一扫,查看更多资料

Chapter 13　Ceramics Matrix Composite

13.1　Introduction

A great variety of silicate matrices have been considered for the fabrication of fiber-reinforced glass and glass-ceramic matrix composites. Typical matrices investigated are listed in Table 13.1. Table 13.2 gives an overview of different composite systems developed and some of the most remarkable properties achieved.

Table 13.1　Some glass and glass-ceramic matrices commonly used to fabricate fiber-reinforced composites

Material	Elastic modulus / GPa	Thermal expansion coefficient ($\times 10^{-6}$/℃)	Fracture strength / MPa	Density (g/cm^3)	Softening point / ℃
Glass matrices					
Silica glass	84	1.8	70~105	2.5	1 300
Borosilicate	63	3.3	70~100	2.2	815
Aluminium silicate	90	4.1	80	2.6	950
Glass-ceramic matrices					
Lithium aluminosilicate	100	1.5	100~150	2.0	n. a.
Cordierite (magnesium aluminosilicate)	119	1.5~2.5	110~170	2.6~2.8	1 450
Barium magnesium aluminosilicate	125	1.2	140	2.8	1 450
Calcium aluminosilicate	110	n. a.	100~130	3.0	1 550

(1) Carbon fiber-reinforced glass matrix composites

The first paper on the fabrication and characterization of carbon fiber-reinforced glass matrix composites was published in 1969. Major developments in these composite systems were carried out during the 70s and 80s, especially in USA, England and Germany. There are still considerable research efforts worldwide in this area. A recent development is the use of nitride glass matrices, for example Y-Si-Al-O-N glass.

The interfacial properties of carbon fiber composites depend primarily on the physical structure and chemical bonding at the interface and on the type of carbon fiber used. Moreover, the interfacial strength in these composites can be influenced by changing the chemistry of the matrix. The major disadvantage of these composites is the limited temperature capability in oxidising atmospheres at high temperature. The oxidation behaviour under different conditions has been investigated.

(2) Nicalon and Tyranno fiber-reinforced glass and glass-ceramic matrix composites

Glass and glass-ceramic matrix composites reinforced by SiC-based fiber of the type Nicalon or Tyranno combine strength and toughness with the potential for high temperature oxidation resistance. A great variety of silicate matrices have been reinforced by these fibers (see also Table 13.2). Several products have reached commercial exploitation[①], such as the material Fortadur and the composite Tyrannohex. Glass matrices such as silica, borosilicate and aluminosilicate have been used, as well as glass-ceramic matrices, such as lithium aluminosilicate (LAS), magnesium aluminosilicate (MAS), calcium aluminosilicate (CAS), barium aluminosilicate (BAS), barium magnesium aluminosilicate (BMAS), calcium magnesium aluminosilicate (CMAS), yttrium aluminosilicate (YAS), lithium magnesium aluminosilicate (LMAS) and yttrium magnesium aluminosilicate (YMAS). Oxynitride glass matrices (e.g., $Y_{44}Si_{81}Al_{48}O_{24}N_{40}$ and $Li_{40}Si_{80}Al_{40}O_{216}N_{16}$ compositions) have also been considered. The use of refractory glass-ceramic matrices has the objective to develop materials with high temperature capability (>1 000 ℃).

Table 13.2 Overview of fiber-reinforced glass-ceramic matrix composites

Matrix/fiber	Properties investigated
Borosilicate/SiC (Nicalon)	$\sigma = 1\,200$ MPa, $K_{Ic} = 18$ MPm$^{1/2}$ (up to 600 ℃)
Magnesium aluminosolicate	$\sigma = 60$ MPa (up to 600 ℃)
SiC-monofilament	$\sigma = 70$ MPa (up to 1 100 ℃ in air)
Lithium aluminosolicate/SiC-(Nicalon)	$\sigma = 1\,100$ MPa (up to 1 100 ℃ in argon)
Silica/carbon	$\sigma = 800$ MPa (at room temperature)
Borosilicate/metal (Ni-Si-B alloy)	$\sigma = 225$ MPa (up to 500 ℃)
Borosilicate/SiC (Nicalon)	$\sigma = 840$ MPa, $K_{Ic} = 25$ MPm$^{1/2}$ (up to 530 ℃)
Aluminosilicate/SiC (Nicalon)	$\sigma = 1\,200$ MPa, $K_{Ic} = 36$ MPm$^{1/2}$ (up to 700 ℃)
Borosilicate/mullite	$\sigma = 150$ MPa, $K_{Ic} = 2.5$ MPm$^{1/2}$ (up to 600 ℃)

Chapter 13 Ceramics Matrix Composite

	(to be continued)
Matrix/fiber	Properties investigated
Silica/SiC (Nicalon), 2-D	σ = 205 MPa (up to 1 000 ℃)
Magnesum aluminosilicate/SiC (Nicalon)	σ = 1 057 MPa (up to 500 ℃) σ = 414 MPa (up to 700 ℃)
Barium-magnesium luminosicate/SiC (Nicalon)	σ = 900 MPa (up to 1 100 ℃)

Notes: σ: ultimate fracture strength, K_{Ic}: fracture toughness, 2-D: 2 dimensional reinforcement

A distinct member of the family of oxycarbide fiber materials is Tyrannohex, a composite developed on the basis of bonded Tyranno fibers. The fiber content reaches values up to 90 vol %. Due to the near absence of a matrix phase, the material retains high strength up to very high temperatures (195 MPa at 1 500 ℃) and it has a very high creep resistance. The new generation Tyrannohex material fabricated with novel sintered high-performance SA-Tyranno fiber has shown improved thermomechanical properties up to 1 600 ℃.

Detailed high-resolution electron microscopy and microanalytical investigations have been conducted to elucidate the phase structure and microchemical composition of the fiber/matrix interfacial region in a variety of composite systems. These investigations showed that the interfacial zone is occupied by a carbon-rich thin layer of thickness 10~50 nm, depending on the matrix/fiber combination. The carbon-rich interfacial layer is clearly observed. High-resolution[②] electron microscopy of these interfaces has revealed that the layers contain graphitic carbon textured to varying degrees. This carbonaceous layer is weaker than the matrix so that the fibers are effective in deflecting matrix cracks and promote fiber pull-out during composite failure. As explained below, this is the main mechanism leading to the high fracture toughness and flaw-tolerant fracture behaviour of this class of composites. At the same time, the interface layer allows load transfer from matrix to fiber so that strengthening takes place.

The mechanisms of formation of this carbon-rich layer at the fiber-matrix interfaces have been studied. It has been proposed that the carbon layer is the result of a fortunate combination of silicate matrix chemistry and nonstoichiometric/noncrystalline fiber structure. The solid-state reaction between the SiC in the fiber and oxygen from the glass and the fiber surface can be written as

$$SiC(s) + O_2 \longrightarrow SiO(s) + C(s)$$

When using SiC-based fibers of the type Nicalon R and Tyranno R, the temperature capability of the composites depends strongly on the environment. In oxidising atmospheres (i.e., hot air), the stability of the carbon-rich fiber-matrix interface represents the limiting factor. One way to alleviate the problem of degradation of the carbon layer is to use protective coatings on the fibers. These coatings should act as a diffusion barrier layer and a low decohesion layer. A variety of single-layer and two-layer coatings produced by chemical vapour deposition (CVD) have been tried in attempts to replicate the mechanical response of the carbon-rich interfaces, including C, BN, BN(+C), BN/SiC and C/SiC coatings.

(3) SiC and boron monofilament-reinforced glass matrix composites

Monofilament SiC fibers, produced by chemical vapour deposition (CVD), are normally fabricated in diameters ranging from 100 to 140 μm. Since these monofilaments are less flexible than the fibers derived from organo-silicon polymers, only simple shaped components can be produced. Composites were fabricated using up to 65 vol % SiC monofilaments (type SCS-6) in a borosilicate glass matrix. The higher elastic modulus of SiC monofilament results in composites significantly stiffer than those fabricated using Nicalon or Tyranno fibers. A major disadvantage of using these filaments is their large diameter, which may lead to extensive microcracking in the matrix and, therefore, to unacceptable low off-axis[③] strength.

SiC monofilaments have been used in borosilicate and aluminoslicate glass matrices for the fabrication of model composites, including composites with a transparent matrix. These were conveniently used to investigate the development of matrix microcracking under flexure stresses since the formation of the microcracking pattern in the transparent glass matrix could be observed in situ.

Researchers at NASA Lewis Research Centre have used SiC monofilaments for the reinforcement of glass-ceramic matrices of refractory compositions, including strontium aluminosilicate (SAS) and barium aluminosilicate (BAS), with temperature capability of up to 1 600 ℃.

Boron monofilaments (about 20 vol %) were used by Tredway and Prewo to reinforce borosilicate glass. They also introduced carbon fiber yarn to fill in the glass-rich regions between monofilaments in order to toughen the matrix and provide additional structural integrity of the composite.

(4) Glass/glass-ceramic fiber-glass matrix composite

The first report on preparation of this class of composites was published by Japanese researchers using chopped SiCaON glass fibers and matrices. The fabrication of model composites consisting of a single optical fiber embedded in borosilicate glass and of continuous oxynitride glass fiber reinforced glass matrix composites with a SiO_2-B_2O_3-La_2O_3 glass matrix has been also reported. Further work on silicate glass matrix composites with continuous silicate glass fiber reinforcement has been conducted in Germany and the UK. In a similar way as when using crystalline fibers, there is need for engineering the interface in glass/glass composites in order to avoid strong bonding which was resulted from chemical reactions during composite fabrication. A declared goal of further research in the area of glass/glass composites is the development of transparent or translucent materials showing high fracture toughness and adequate flaw tolerant behaviour. A major challenge in the development of such composites is to be able to incorporate interfaces which are optically, chemically, thermally and mechanically compatible with the matrix and fibers. One suggested way to tailor the interfaces is to use dense, transparent (or translucent) nanosized oxide coatings. In this regard, translucent tin dioxide may be a good candidate. Other suggested oxide coating on silicate fibers used in transparent glass matrix composites is titanium dioxide produced by the sol-gel method④. Glass fiber-reinforced glass matrix composite materials may lead to interesting products for replacement of laminate glass in applications requiring relatively high fracture.

(5) Metal fiber-reinforced glass matrix composites

An advantage of these composites is the increased resistance to fiber damage during composite processing which results from the intrinsic ductility of metallic fibers and the possibility of exploiting their plastic deformation for composite toughness enhancement. A penalty is paid due to the relatively low thermal capability and poor chemical resistance of the metallic fiber reinforcement, limiting the application temperature and environment. In earlier study, Ducheyne and Hench fabricated composite materials with a bioactive glass (Bioglass) matrix and stainless steel fiber reinforcement by an immersion technique. Donald. et al. have reported on the fabrication of glass and glass-ceramic matrix composites reinforced by stainless steel and Ni-based alloy filaments with diameters in the range of 4 to 22 μm. Glass-encapsulated metal filaments prepared by the Taylor-wire process were used for the fabrication of the composites.

Russian researchers have demonstrated the use of 2-dimensional metal fiber structures to reinforce glass and glass-ceramics. However, only a limited number of glass matrices and metallic fiber reinforcements were tried. More recently, Boccaccini and co-workers have used electrophoretic deposition to fabricate a number of glass matrix composites containing 2-dimensional metal fiber reinforcement. In particular, soda-lime⑤, borosilicate, cathode-ray tube⑥ recycled glass and bioactive glass were used as matrices and a variety of commercially available stainless steel fiber mats were used as reinforcement. The fracture surface exhibits fiber pull-out and partial plastic deformation of the fibers, which indicates flaw-tolerant behavior of the composite.

13.2 Processing of Ceramic Matrix Composites

The easy processing in comparison to polycrystalline ceramic matrix composites is one of the outstanding attributes of glass and glass-ceramic matrix composites. This is due to the ability of glass to flow at high temperatures in a similar way to resins, which is exploited in different fabrication strategies as discussed below.

(1) Slurry infiltration and hot-pressing

Hot-pressing of infiltrated fiber tapes or fabric lay-ups are the most extensively used technique to fabricate dense fiber-reinforced glass matrix composites. Figure 13.1 shows a schematic diagram of the standard fabrication process. For proper densification, the time-temperature-pressure schedule during hot-pressing must be optimized so that the glass viscosity is low enough to permit the glass to flow into the spaces between individual fibers within the tows. The processing temperature must be chosen after taking into account the possible occurrence of crystallization of the glass matrix and the "in-situ" formation of the carbonaceous interface, when oxycarbide fibers (e. g., Nicalon or Tyranno) are used. The pressure (of about 10~20 MPa) is usually applied after the temperature reaches the softening point of the matrix glass. In this manner almost fully dense composites can be fabricated.

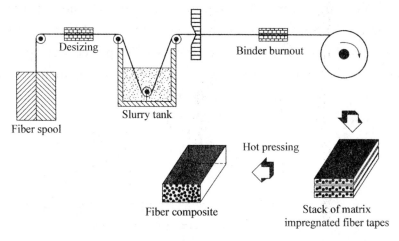

Figure 13.1 **The standard slurry impregnation and hot-pressing route to fabricate fiber-reinforced glass matrix composites**

When a glass-ceramic matrix is required, the densified composites are subjected to a "ceraming" heat-treatment after densification. In this case, an optimize time-temperature "window" must be found where densification takes place by viscous flow of the glass matrix before the onset of crystallization. Ideally, the temperature range at which densification occurs at maximum rate lies between the softening temperature of the glass and the onset of crystallization temperature. A post-densification heat-treatment leads to the desired crystalline, refractory microstructure of the matrix. In some systems, the crystallization of the glass-ceramic matrix can be achieved during hot-pressing. The possibility for net shape fabrication of composite components using hot-pressing has been shown in the literature.

(2) Tape casting

The equipment used for casting green sheets for use in LTCCs is shown in Figure 13.2. Currently, a variety of casting equipments are being manufactured but in general, the equipment consists of a carrier film conveyor, casting head, slurry dispenser, drying area and sheet take-up unit. The carrier film conveyor fulfills the role of conveying the plastic carrier film, fed from a roll, to the casting head. Since the plastic film is the carrier of the cast sheet, it is desirable that it has no wrinkles, and travels in a straight line at an even speed. At the casting head, the ceramic slurry is dispensed onto the carrier film. The slurry dispenser is for volumetric feed of the slurry to the casting head in order to produce the ceramic green sheet reliably and continuously. The drying area drives off the solvent in the cast ceramic slurry to produce a dried sheet. Drying normally uses infrared heaters or hot air. The drying temperature profile is adjusted taking into account the drying rate of the slurry and the speed of the carrier film. The sheet take-up unit picks up the dried ceramic green sheet in a roll. Some take-up units remove the green sheet from the carrier film while others take up the carrier film as well. PET (polyethylene terephthalate) film is commonly used for carrier film, and according to requirements, a silicone release agent is applied in order to improve peelability.

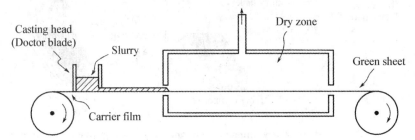

Figure 13.2 Conceptual diagram of green sheet casting equipment

(3) Sol-gel, colloidal routes and electrophoretic deposition

Using the sol-gel approach, matrices are produced from metal alkoxide solutions or colloidal sols as precursors. A great variety of glass and glass-ceramic matrices have been prepared by sol-gel processing, including borosilicate, and lithium (LAS), magnesium (IVIAS), barium (BAS), calcium (CAS) and sodium (NAS) aluminosilicates. The composites are fabricated by drawing the fibers through a sol in order to deposit a gel layer on the fiber surfaces. After gelation and drying, the prepregs are densified by hot-pressing, but pressureless sintering densification is also possible. Potential advantages of the sol-gel method over slurry processing include: more effective fiber infiltration enabling a reduction of the hot-pressing temperature, which in turn leads to limitation of damage to the fibers, and the ability to tailor matrix composition in order to control thermal expansion mismatch and fiber-matrix interfacial chemistry. Another advantage of using sol-gel processing is the possibility of infiltrating complex fiber architectures and to develop nearest size and shape manufacturing technologies. Disadvantages of the sol-gel method are associated with large matrix shrinkage during drying and densification due to low solid content of the sols, leading to matrix cracking and residual porosity. Moreover, the process is time-consuming, usually requiring; a large number of infiltration/drying steps.

Another way to fabricate glass matrix composites is to combine the sol-gel with slurry approaches. This can overcome the large shrinkages inherent in the sol-gel process while maintaining the advantages of ease of infiltration and lower fabrication temperatures. An alternative process to fabricate glass and glass-ceramic matrix composites involves electrophoretic deposition (EPD) to infiltrate the fiber preform with matrix material followed by conventional pressureless sintering or hot-pressing for densification of the composite. A schematic diagram of the electrophoretic deposition cell is shown in Figure 13.3. If the deposition electrode is replaced by a conducting fiber preform, the suspended charged (nano) particles will be attracted into and deposited within it. Using EPD, a range of glass and glass-ceramic matrix composites containing 2-dimensional fiber reinforcement, including composites of tubular shape, have been fabricated.

(4) Powder technology, sintering and hot-pressing

Acritical step in this fabrication approach is the mixing of the glass matrix powders and the reinforcing elements. The homogeneous distribution of these in the glass matrix is a fundamental requirement for obtaining high-quality composites with optimized properties. Inhomogeneous distribution of inclusions, forming of particle clumps or agglomerates or inclusion-inclusion interactions, may lead to microstructural defects such as pores and cracks, which will have a negative effect on the mechanical properties of the products. The improvement of the mixing techniques includes optimizing the particle sizes and the size distributions of the

Chapter 13 Ceramics Matrix Composite

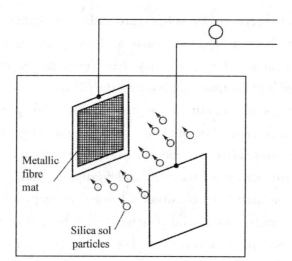

Figure 13.3 A schematic diagram of the electrophoretic deposition cell

powders and the use of the wet-mixing routes, i. e., mixing in water or isopropanol with the addition of binders such as PVA coupled with ultrasonic or magnetic stirring. Adequate mixing of the powder matrix and the reinforcing elements is particularly problematic when chopped fibers are used, and careful control of the processing variables, including slurry viscosity, fiber content and stirring velocity is required. A fairly homogeneous glass powder-chopped fiber mixture has been achieved, which could not have been reached by dry-mixing techniques. Another approach to improve homogeneity of the mixture is the coating of the reinforcing elements by a thin layer of the matrix material. This approach presupposes the development of a technique to synthesis submicrometric glass particles.

The most economical process involves densification of the green bodies by simple pressureless sintering at temperatures between the glass transition and the melting temperatures of the glass matrix. Sintering in glass matrices occurs by a viscous flow mechanism. During the sintering, reactions at the inclusion/matrix interfaces or degradation of the inclusions may take place, especially when the inclusions are nonoxide ceramics (e. g., SiC) or metallic. If the glass matrix used is prone to crystallization or the aim is to produce a glass-ceramic matrix, the sintering procedure must be optimized in order to avoid the onset of crystallization before the densification by viscous flow has been fully completed. One way to achieve this is by increasing the heating rate during sintering in order to delay the nucleation and growth of crystalline phases. Thus an ideal heating schedule to produce glass-ceramic matrix composites should include three independent stages: densification, nucleation and crystallization.

The presence of inclusions can affect the formation of crystalline phases in a glass matrix. For example, it has been shown that aluminum-containing ceramic inclusions, e. g., alumina, mullite or aluminum nitride, may suppress cristobalite formation in borosilicate glass during sintering. In general, the presence of rigid inclusions will jeopardize the densification process driven by viscous flow sintering, and a problem that has been well studied both experimentally and theoretically. In practice, the maximum volume fraction of inclusions is suitable to yield high-quality, and dense composites by pressureless sintering are about 15 vol %. For higher contents, the consolidation of the composite powder mixtures is usually conducted by hot-pressing. This technique involves the sintering of the composite glass powder mixture under uniaxial pressure, usually in the range of 5~20 MPa, in a die. While allowing the fabrication of composites with a high volume fraction of inclusions (of up to 90 vol %) and without porosity, this technique is cost-intensive[①] and has limitations regarding the shape and complexity of the parts that can be produced. The fairly homogeneous distribution of the reinforcing elements and the absence of porosity, cracks or other microstructural defects are evident, which indicates improved mechanical properties in these composites.

(5) Other fabrication methods

A technique suggested to fabricate complex shaped structures is the matrix transfer molding technique. Woven structures used as reinforcement are arranged inside a mold cavity. Fluid matrix is transferred at high temperature into the mold cavity to fill the void space around the reinforcement structure. In this way, for example, thin-walled cylinders can be fabricated. Another proposed method for obtaining structures of complex shapes uses a superplastically deformed foil. The foil is used to partially encapsulate the composite sample and a part holding die, which are both contained in a rigid box held in a press. This method allows near-net-shape manufacturing but it involves complex and time consuming operations.

Efforts have also been made to evaluate the utility of the polymer precursor method for processing fiber reinforced glass-ceramics. In general, the precursor approach provides access to glass-ceramics via low temperature processing methods with good control of chemical and phase homogeneity. Because the polymer precursor method has the potential to yield near-net-shape products at relatively low temperatures, this processing route represents an interesting alternative to the established slurry and hot-pressing technique.

Another simple method of fabricating unidirectional fiber reinforced glass matrix composite rods is based on pultrusion. The technique is simple and can potentially be used for a wide range of fibers and matrices, and it enables

Chapter 13 Ceramics Matrix Composite

unidirectional composites in the form of rods with different cross-sections to be fabricated. A related processing method yielding composite shapes in the form of longitudinal rods or cylinders is the extrusion technique developed by Klein and Roeder.

The densification of Nicalon fiber-reinforced borosilicate glass matrix composites by microwave heating has also been investigated. The results demonstrated that microwave processing could be a highly efficient method, saving time and reducing costs, in comparison with the traditional hot-pressing densification.

◆ Notes

① commercial exploitation 商业炒作；商业利用；商业开发
② high-resolution 高分辨率
③ off-axis 离轴；偏轴；轴外
④ sol-gel method 溶胶—凝胶法
⑤ soda-lime 碱石灰
⑥ cathode-ray tube 阴极射线管
⑦ cost-intensive 成本密集

◆ Vocabulary

alkoxide [æl'kɒksaɪd] n.［有化］醇盐；酚盐
aluminosilicate [əˌljuːmɪnəʊ'sɪlɪkət] n.［无化］铝硅酸盐
architecture ['ɑːkɪtektʃə] n. 建筑学；建筑风格；建筑式样；架构
borosilicate [ˌbɔːrəʊ'sɪlɪkeɪt] n. 硼硅酸盐
carbonaceous [ˌkɑːbə'neɪʃəs] adj.［植］碳质的；碳的，含碳的
colloidal [kə'lɔɪdəl] adj. 胶体的；胶质的；胶状的
conveyor [kən'veɪə] n. 输送机，［机］传送机；传送带；运送者，传播者
cordierite ['kɔːdɪəraɪt] n.［矿物］堇青石
cristobalite [krɪs'təʊbəlaɪt] n. 方石英；方晶石
decohesion [ˌdiːkəʊ'hiːʒən] n. 检波器恢复常态；减聚力

densification [ˌdensɪfɪ'keɪʃən] n. 密实化；封严；［化学］稠化
dispenser [dɪ'spensə] n. 药剂师；施与者；分配者；自动售货机
exploitation [ˌeksplɔɪ'teɪʃ(ə)n] n. 开发，开采；利用；广告推销；剥削
immersion [ɪ'mɜːʃ(ə)n] n. 沉浸；陷入；专心
isopropanol [ˌaɪsə'prəʊpənɒl] n.［有化］异丙醇
jeopardize ['dʒepədaɪz] vt. 危害；使陷危地；使受危困
microanalytical [ˌmaɪkrəʊnə'lɪtɪkl] adj. 微量分析的
microcracking [maɪkrɪk'rækɪŋ] n. 微裂缝；显微裂纹
nicalon [nɪ'kælən] n. 碳化硅
organo ['ɔːgənəʊ] adj. 有机金属的
oxycarbide [ɒksɪ'kɑːbaɪd] n. 碳氧化物

oxynitride [ɒksɪ'naɪtraɪd] n. 氮氧化物
problematic [prɒblə'mætɪk] adj. 问题的;有疑问的;不确定的
prone [prəʊn] adj. 俯卧的;有……倾向的,易于……的
pultrusion [pʌl'truːʒn] n. 拉挤成型;挤压成型
resolution [rezə'luːʃ(ə)n] n. [物]分辨率;决议;解决;决心
silicone ['sɪlɪkəʊn] n. 硅树脂;[口腔]硅酮

slurry ['slʌrɪ] n. 泥浆;悬浮液
sol [sɒl] n. 溶胶
translucent [træns'luːs(ə)nt] adj. 透明的;半透明的
transparent [træn'spær(ə)nt] adj. 透明的;显然的;坦率的;易懂的
unidirectional [ˌjuːnɪdɪ'rekʃ(ə)n(ə)l] adj. 单向的;单向性的
woven ['wəʊvn] n. 机织织物 v. 编织;交织(weave 的过去分词);编造 adj. 织物的

Exercises

1. Translate the following Chinese phrases into English

(1) 纤维增强玻璃　　(2) 界面性质　　(3) 温度性能

(4) 氧化行为　　(5) 断裂韧性　　(6) 基质相

(7) 半透明材料　　(8) 金属纤维　　(9) 料浆浸渗

(10) 非氧化物陶瓷

2. Translate the following English phrases into Chinese

(1) glass-ceramic matrix composites　　(2) chemical bonding

(3) oxidising atmospheres　　(4) SiC-based fiber

(5) ultimate fracture strength　　(6) two-layer coatings

(7) extensive microcracking　　(8) immersion technique

(9) electrophoretic deposition　　(10) softening temperature

(11) mold cavity　　(12) thin-walled cylinders

3. Translate the following Chinese sentences into English

(1) $CaCO_3$ 或石灰石存在于多种物质中。

(2) 由铝和碳化硅等材料构成的纤维有益于高温下的应用,与此同时环境上的应战也成为一个议题。

(3) 利用压力将气泡挤出,然后给模子加热,使基质变成固体。

(4) 目前,被归类为先进陶瓷的高温陶瓷超导体不占有主要市场。

(5) 居里温度以上电阻的急剧增加,对于热敏电阻的实际应用及导电机制的基础研究来说都是非常令人激动的发现。

(6) 在电性能和机械强度方面都在进行改进和开发,例如陶瓷固体电解质和电极材料的改进。

(7) 无机化合物作为陶瓷基复合材料(CMCs)的一种特殊组分,由于它们具有显著的热力学特性和耐高温性能(可在 1 200℃使用),得到了人们的极大关注。

Chapter 13 Ceramics Matrix Composite

4. Translate the following English sentences into Chinese

(1) Ceramic compounds can be defined as inorganic-compounds made by heating clay or other mineral matter a high temperature at which they partially melt and bond together.

(2) Preforms are made of particulate, which can be produced by any of the ceramic green-body-forming methods, such as pressing, isostatic pressing, slip casting, extrusion or injection molding.

(3) Since the reinforcement material is of primary importance in the strengthening mechanism of a composite, it is convenient to classify composites according to the characteristics of the reinforcement.

(4) Since the thermal resistance and mechanical properties of ceramic polycrystalline materials compare very favorably with those of alumina and yttria, they are ideal for use in high-temperature viewports where thermal resistance and a high degree of transparency are both required.

(5) With the availability of continuous ceramic fibers, there has been a tremendous interest in the development of ceramic matrix composite (CMC) possessing high temperature capability (T > 1 500 ℃), noncatastrophic failure mechanisms and low density.

(6) Oxide-CMCs would obviously be the best choice, from a thermodynamic standpoint, for long term applications in oxidizing atmospheres.

5. Translate the following Chinese essay into English

复合材料按用途主要可分为结构复合材料和功能复合材料。结构复合材料是指主要作为承力结构使用的材料,功能材料是指除力学性能以外还提供其他物理、化学、生物等性能的复合材料。

6. Translate the following English essay into Chinese

Ceramic processing can be described simply as consisting of two steps: (1) a cold step, in which the ceramic part is formed or shaped into a "green" part or perform, and (2) a hot, in which the green compact is first subjected to heat to dry up any liquid phase formed during the processing and then subjected to higher heating, known as firing, sintering or densification.

扫一扫,查看更多资料

Chapter 14　Polymer Matrix Composite

14.1　Introduction

In its most basic form a composite material is one which is composed of at least two elements working together to produce material properties that are different to the properties of those elements on their own. In practice, most composites consist of a bulk material (the matrix), and a reinforcement of some kind, added primarily to increase the strength and stiffness of the matrix. This reinforcement is usually in fibre form. In the most common man-made composites, polymer matrix composites (PMCs) are the most common and will be discussed here. Also known as FRP-fibre reinforced polymers (or plastics), these materials use a polymer-based resin as the matrix, and a variety of fibres such as glass, carbon and aramid as the reinforcement.

Resin systems such as epoxies and polyesters have limited use for the manufacture of structures on their own, since their mechanical properties are not very high when compared to, for example, most metals. However, they have desirable properties, most notably their ability to be easily formed into complex shapes. Materials such as glass, aramid and boron have extremely high tensile and compressive strength lout in solid form these properties are not readily apparent. This is due to the fact that when stressed, random surface flaws will cause each material to crack and fail well below its theoretical breaking point[1]. To overcome this problem, the material is produced in fibre form, so that, although the same number of random flaws will occur, they will be restricted to a small number of fibres with the remainder exhibiting the material's theoretical strength. Therefore a bundle of fibres will reflect more accurately the optimum performance of the material. However, fibres alone can only exhibit tensile properties along the fibre length, in the same way as fibres in a rope.

It is when the resin systems are combined with reinforcing fibres such as glass, carbon and aramid, those exceptional properties can be obtained. The resin matrix spreads the load applied to the composite between each of the individual fibres and also protects the fibres from damage caused by abrasion and impact. High strengths and stiffnesses, ease of moulding complex shapes, high environmental resistance all coupled with low densities, make the resultant composite superior to metals for many applications.

Since PMCs combine a resin system and reinforcing fibres, the properties of the resulting composite material will combine something of the properties of the resin on its own with that of the fibres on their own (see Figure 14.1).

Overall, the properties of the composite are determined by:

① The properties of the fibre.

② The properties of the resin.

③ The ratio of fibre to resin in the composite (fibre volume fraction).

④ The geometry and orientation of the fibres in the composite.

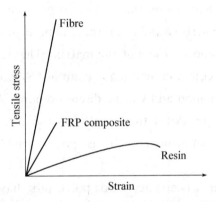

Figure 14.1 Schematic tensile stress-strain diagram of fibre, FRP composite and resin

The first two will be dealt with in more detail later. The ratio of the fibre to resin derives largely from the manufacturing process used to combine resin with fibre, as will be described in the section on manufacturing processes. However, it is also influenced by the type of resin system used, and the form in which the fibres are incorporated. In general, since the mechanical properties of fibres are much higher than those of resins, the higher the fibre volume fraction the higher will be the mechanical properties of the resultant composite. In practice there are limits to this, since the fibres need to be fully coated in resin to be effective, and there will be an optimum packing of the generally circular cross-section fibres. In addition, the manufacturing process used to combine fibre with resin leads to varying amounts of imperfections and air inclusions.

Typically, with a common hand lay-up process as widely used in the boat-building[②] industry, a limit for FVF is approximately 30~40 %. With the higher quality, more sophisticated and precise processes used in the aerospace industry, FVF's approaching 70 % can be successfully obtained.

The geometry of the fibres in a composite is also important since fibres have their highest mechanical properties along their lengths, rather than across their widths. This leads to the highly anisotropic properties of composites, where, unlike metals, the mechanical properties of the composite are likely to be very different when tested in different directions. This means that it is very important when considering the use of composites to understand at the design stage, both the magnitude and the direction of the applied loads. When correctly accounted for, these anisotropic properties can be very advantageous since it is only necessary to put material where loads will be applied, and thus redundant material is avoided.

It is also important to note that with metals the properties of the materials are largely determined by the material supplier, and the person who fabricates the materials into a finished structure can do almost nothing to change those in built properties. However, a composite material is formed at the same time as the structure is itself being fabricated. This means that the person who is making the structure is creating the properties of the resultant composite material, and so the manufacturing processes they use have an unusually critical part to play in determining the performance of the resultant structure.

(1) Loading

There are four main direct loads that any material in a structure has to withstand: tension, compression, shear and flexure.

(2) Tension

Figure 14.2 shows a tensile load applied to a composite. The response of a composite to tensile loads is very dependent on the tensile stiffness and strength properties of the reinforcement fibres, since these are far higher than the resin system on its own.

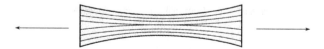

Figure 14.2 Tensile load applied to a composite

(3) Compression

Figure 14.3 shows a composite under a compressive load. Here, the adhesive and stiffness properties of the resin system are crucial, as it is the role of the resin to maintain the fibres as straight columns and to prevent them from buckling.

Figure 14.3 Composite under a compressive load

(4) Shear

Figure 14.4 shows a composite experiencing a shear load. This load is trying to slide adjacent layers of fibres over each other. Under shear loads the resin plays the major role, transferring the stresses across the composite. For the composite to perform well under shear loads the resin element must not only exhibit good mechanical properties must but also have high adhesion to the reinforcement fibre. The interlaminar shear strength of a composite is often used to indicate this property in a multilayer composite (laminate).

Figure 14.4 Composite experiencing a shear load

(5) Flexure

Flexural loads are really a combination of tensile, compression and shear loads. When loaded as shown, the upper face is put into compression, the lower face into tension and the central portion of the laminate experiences shear (Figure 14.5).

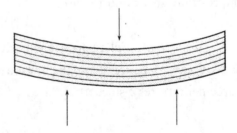

Figure 14.5 Composite under a flexural load

14.2 Fabrication Processes

The mixture of reinforcement/resin does not really become a composite material until the last phase of the fabrication, that is, when the matrix is hardened. After this phase, it would be impossible to modify the material, as in the way one would like to modify the structure of a metal alloy using heat treatment, for example. In the case of polymer matrix composites, this has to be polymerized, for example, polyester resin. During the solidification process, it passes from the liquid state to the solid state by copolymerization with a monomer that is mixed with the resin. The phenomenon leads to hardening. This can be done using either a chemical (accelerator) or heat. The following pages will describe the principal processes for the formation of composite parts.

(1) Molding processes

The flow chart in Figure 14.6 shows the steps found in all molding processes. Forming by molding processes varies depending on the nature of the part, the number of parts and the cost. The mold material can be made of metal, polymer, wood or plaster.

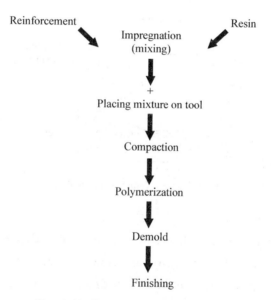

Figure 14.6 Steps in molding process

(a) Contact molding

Contact molding (see Figure 14.7) is open molding (there is only one mold, either male or female). The layers of fibers impregnated with resin (and

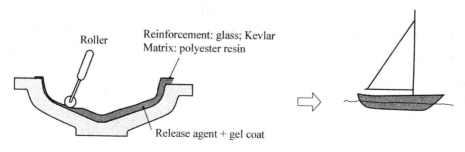

Figure 14.7 Contact molding

accelerator) are placed on the mold. Compaction is done using a roller to squeeze out the air pockets. The duration for resin setting varies, depending on the amount of accelerator, from a few minutes to a few hours. One can also obtain parts of large dimensions at the rate of about 2 to 4 parts per day per mold.

(b) Compression molding

With compression molding (see Figure 14.8), the counter mold will close the mold after the impregnated reinforcements have been placed on the mold. The whole assembly is placed in a press that can apply a pressure of 1 to 2 bars. The polymerization takes place either at ambient temperature or higher. The process is good for average volume production: one can obtain several dozen parts a day (up to 200 with heating). This has application for automotive and aerospace parts.

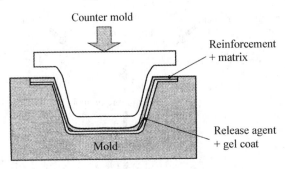

Figure 14.8　Compression molding

(c) Molding with vacuum

This process of molding with vacuum is still called depression molding or bag molding. As in the case of contact molding described previously, one uses an open mold on top of which the impregnated reinforcements are placed. In the case of sandwich materials, the cores are also used. One sheet of soft plastic is used for sealing (this is adhesively bonded to the perimeter of the mold). Vacuum is applied under the piece of plastic (see Figure 14.9). The piece is then compacted due to the action of atmospheric pressure, and the air bubbles are eliminated. Porous fabrics absorb excess resin. The whole material is polymerized by an oven or by an autoclave under pressure (7 bars in the case of carbon/epoxy to obtain better mechanical properties), or with heat, or with electron beam, or X-rays; see Figure 14.10). This process has applications for aircraft structures, with the rate of a few parts per day (2 to 4).

Chapter 14　Polymer Matrix Composite

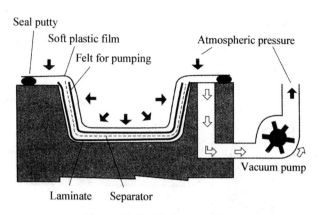

Figure 14.9　Vaccum molding

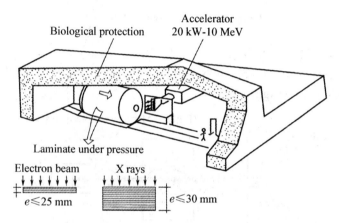

Figure 14.10　Electron beam of X-ray molding

(d) Resin injection molding

With resin injection molding (see Figure 14.11), the reinforcement (mats and fabrics) is put in place between the mold and counter mold. The resin (polyester or phenolic) is injected. The mold pressure is low. This process can produce up to 30 pieces per day. The investment is less costly and has application in automobile bodies.

Figure 14.11　Resin injection molding

(e) Molding by injection of premixed

The process of molding by injection of premixed allows automation of the fabrication cycle (rate of production up to 300 pieces per day).

Chapter 14 Polymer Matrix Composite

① Thermoset resins

It can be used to make components of auto body. The schematic of the process is shown in Figure 14.12.

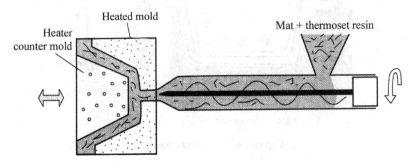

Figure 14.12 Injection of premixed

② Thermoplastic resins

It can be used to make mechanical components with high temperature resistance, as shown in Figure 14.13.

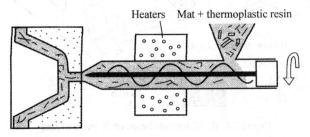

Figure 14.13 Injection of thermoplastic premixed

(f) Molding by foam injection

Molding by foam injection (see Figure 14.14) allows the processing of pieces of fairly large dimensions made of polyurethane foam reinforced with glass fibers. These pieces remain stable over time, with good surface conditions, and have satisfactory mechanical and thermal properties.

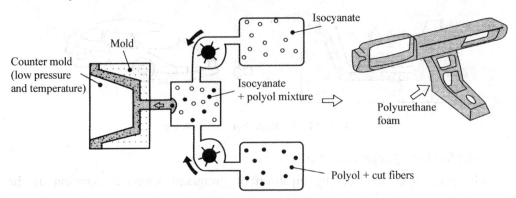

Figure 14.14 Foam injection

(g) Molding of components of revolution

The process of centrifugal molding (see Figure 14.15) is used for the fabrication of tubes. It allows homogeneous distribution of resin with good surface conditions, including the internal surface of the tube. The length of the tube depends on the length of the mold. Rate of production varies with the diameter and length of the tubes (up to 500 kg of composite per day).

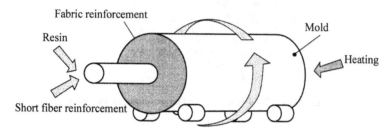

Figure 14.15 Centrifugal molding

The process of filament winding (see Figure 14.16) can be integrated into a continuous chain of production and can fabricate tubes of long length. The rate of production can be up to 500 kg of composite per day. These can be used to make missile tubes, torpillas, containers or tubes for transporting petroleum. For pieces which must revolve around their midpoint, winding can be done on a mandrel. This can then be removed and cured in an autoclave (see Figure 14.17). The fiber volume fraction is high (up to 85%). This process is used to fabricate components of high internal pressure, such as reservoirs and propulsion nozzles.

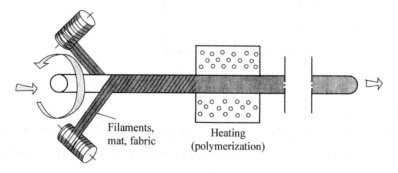

Figure 14.16 Filament winding

Chapter 14 Polymer Matrix Composite

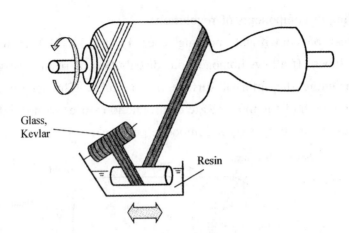

Figure 14.17 Filament winding on complex mandrel

(2) Other forming processes

(a) Sheet forming

This procedure of sheet forming (see Figure 14.18) allows the production of plane or corrugated sheets by corrugation or ribs.

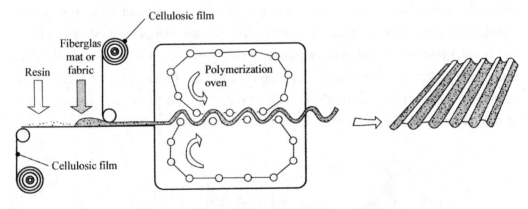

Figure 14.18 Sheet forming

(b) Profile forming

The piece shown in Figure 14.19 is made by pultrusion. This process makes possible the fabrication of continuous open or closed profiles. The fiber content is

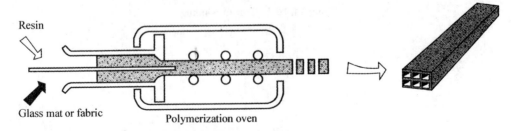

Figure 14.19 Profile forming

important for high mechanical properties. The rate of production varies between 0.5 and 3 m/minute, depending on the nature of the profile.

(c) Stamp forming

Stamp forming (see Figure 14.20) is only applicable to thermoplastic composites. One uses preformed plates, which are heated, stamped and then cooled down.

Figure 14.20 Stamp forming

(d) Preforming by three-dimensional assembly

i) carbon/carbon

The carbon reinforcement is assembled by depositing the woven tows along several directions in space. Subsequently the empty space between the tows is filled by impregnation. The following two techniques are used: impregnation using liquid, pitch is used under a pressure of 1 000 bars, followed by carbonization; impregnation using gas. This involves chemical vapor deposition using a hot gaseous hydrocarbon atmosphere.

ii) silicon/silicon

The reinforcement is composed of filaments of silicon ceramics. The silicon matrix is deposited in the form of liquid solution of colloidal silicon, followed by drying under high pressure and high temperature (2 000 bars, 2 000 ℃). The preforms are then machined. The phases of development of these composites, such as the densification (formation of the matrix) are long and delicate. These make the products very onerous. Applications include missile and launcher nozzles, brake disks, ablative tiles for reentry body of spacecraft into the atmosphere.

(f) Cutting of fabric and trimming of laminates

Some components need a large number of fabric layers (many dozens, can be hundreds). For the small and medium series, it can be quite expensive to operate manually for following the form of a cut respect the orientation specified by the design minimizing waste. There is a tendency to produce a cut or a drape automatically with the following characteristics:

A programmed movement of the cutting machine a rapid cutting machine, such as an orientable vibrating cutting knife or a laser beam with the diameter of about 0.2 mm and a cutting speed varying from 15 to 40 meters/minute, depending on the

Chapter 14　Polymer Matrix Composite

power of the laser and the thickness of the part.

Notes

① breaking point 转换点；断裂点；强度极限
② boat-building 造船

Vocabulary

ablative [ˈæblətɪv] n. 烧蚀材料；离格 adj. 离格的；消融的
autoclave [ˈɔːtə(ʊ)kleɪv] n. 高压灭菌器；高压锅 vt. 用高压锅烹饪 vi. 用高压锅烹饪
carbonization [ˌkɑːbənaɪˈzeɪʃn] n. [化学]碳化；干馏；碳化物
corrugated [ˈkɒrəgeɪtɪd] adj. 波纹的；缩成皱纹的；有瓦楞的 v.（使）起皱纹（corrugate 的过去式）
corrugation [ˌkɔːrʊˈgeɪʃn] n. 起皱；皱状；灌水沟；波纹成形

hydrocarbon [ˌhaɪdrə(ʊ)ˈkɑːb(ə)n] n. [有化]碳氢化合物
phenolic [fɪˈnɒlɪk] n. [胶粘]酚醛树脂 adj. [有化]酚的；[胶粘]酚醛树脂的；石碳酸的
polyurethane [ˌpɒlɪˈjʊərɪθeɪn] n. 聚氨酯；聚亚安酯
propulsion [prəˈpʌlʃ(ə)n] n. 推进；推进力
pultrusion [pʌlˈtruːʒn] n. 拉挤成型；挤压成型
reentry [riːˈentrɪ] n. [航]再入；再进

Exercises

1. Translate the following Chinese phrases into English

（1）人造复合材料　　　　（2）聚合物基树脂　　　　（3）纤维长度
（4）拉伸载荷　　　　　　（5）触压成型　　　　　　（6）制造周期

2. Translate the following English phrases into Chinese

（1）polymer matrix composites　　　（2）environmental resistance
（3）anisotropic property　　　　　　（4）compressive load
（5）injection molding　　　　　　　（6）sheet forming

3. Translate the following Chinese sentences into English

（1）例如，塑料牛奶盒是由一种被称为聚乙烯的聚合物材料制成的，该聚合物具有很高的分子质量。

（2）在热压过程中，在适宜的温度和压力条件下，取向的聚合物纤维被压缩到取向的聚合物片层中，这是一个单组分的方法。

（3）材料可由它们的增强纤维结构和基体分布或增强模式而分类。

（4）混合丝束的生产涉及连续混合增强纤维和基质，织成面料具有良好的悬垂性能。

(5) 玻璃化转变温度 T_g,尽管它并不代表热力学相变,但在很多方面类似于晶体的熔点。

(6) 树脂的化学组成、物理性质影响复合材料的工艺、加工及极限性能。

4. Translate the following English sentences into Chinese

(1) Some polymers, such as nylon and polyesters, can also be formed into fibers that, like hair, are very long relative to their cross-sectional area and are not elastic

(2) Only after many years of research were the right conditions identified for manufacturing this commercially useful polymer.

(3) The greater the number of cross-links in a polymer, the most rigid the material will be.

(4) Co-extrusion technology uses the co-extrusion of two types of polymer tapes of different melting temperatures, cold drawing of the tapes to increase the mechanical properties, and finally consolidation of the tapes.

(5) Unlike thermosets, which undergo a chemical curing reaction for geometry stabilization during forming, the forming process for thermoplastic composites is relatively simple and involves purely physical changes to the material.

(6) Compared to epoxies, polyesters process more easily and are much tougher, whereas phenolics are more difficult to process and brittle, but have higher service temperatures.

(7) Variations in the composition, physical state, or morphology of a resin and the presence of impurities or contaminants in a resin may affect handle ability and process ability, lamina/laminate properties, and composite material performance and long-term durability.

5. Translate the following Chinese essay into English

广义的复合材料是由两种或两种以上的材料复合而成并能赋予其一个独特的综合性能的材料。可以包括金属合金、共聚物、矿产和木材。高分子复合材料的不同之处在于组成材料可以达到分子尺度的复合,而且是机械可分离的。各成分以本体形式构成材料,而且在材料中它们还能保持原来的形状。复合材料的最终性能优于组成材料的性能。

6. Translate the following English essay into Chinese

Polymeric nanocomposites can be considered as an important category of organicinorganic hybrid materials, in which inorganic nanoscale building blocks (e. g., nanoparticles, nanotubes or nanometer thick sheets) are dispersed in an organic polymer matrix. They represent the current trend in developing novel nanostructured materials. When compared to conventional composites based on micrometer-sized fillers, the interface between the filler particles and the matrix in

Chapter 14　Polymer Matrix Composite

a polymer nanocomposite constitutes a much greater area within the bulk material, and hence influences the composite's properties to a much greater extent, even at a rather low filler loading.

扫一扫,查看更多资料

Chapter 15 Nanocomposite

15.1 Introduction

A nanocomposite is as a multiphase solid material where one of the phases has one, two or three dimensions of less than 100 nanometers (nm), or structures having nanoscale repeat distances between the different phases that make up the material. In the broadest sense this definition can include porous media, colloids, gels and copolymers, but is more usually taken to mean the solid combination of a bulk matrix and nanodimensional phase(s) differing in properties due to dissimilarities in structure and chemistry. The mechanical, electrical, thermal, optical, electrochemical, catalytic properties of the nanocomposite will differ markedly from that of the component materials. Size limits for these effects have been proposed, $<$5 nm for catalytic activity, $<$20 nm for making a hard magnetic material soft, $<$50 nm for refractive index changes, and $<$100 nm for achieving superparamagnetism, mechanical strengthening or restricting matrix dislocation movement.

Nanocomposites are found in nature, for example in the structure of the abalon shell and bone. The use of nanoparticle-rich materials long predates the understanding of the physical and chemical nature of these materials. Jose Yacaman, et al, investigated the origin of the depth of colour and the resistance to acids and biocorrosion of Maya blue paint, attributing it to a nanoparticle mechanism. From the mid 1950s nanoscale organo-clays have been used to control flow of polymer solutions (e. g., as paint viscosifiers) or the constitution of gels (e. g., as a thickening substance in cosmetics, keeping the preparations in homogeneous form). By the 1970s polymer/clay composites were the topic of textbooks, although the term "nanocomposites" was not in common use.

In mechanical terms, nanocomposites differ from conventional composite materials due to the exceptionally high surface to volume ratio of the reinforcing phase and/or its exceptionally high aspect ratio. The reinforcing material can be made up of particles (*e. g.*, minerals), sheets (*e. g.*, exfoliated clay stacks) or fibres (*e. g.*, carbon nanotubes or electrospun fibres). The area of the interface between the matrix and reinforcement phase(s) is typically an order of magnitude greater than for conventional composite materials. The matrix material properties are sign/ficantly affected in the vicinity of the reirforcement. Ajayan, *et al.* note that with polymer nanocomposites, properties related to local chemistry, degree of thermoset cure, polymer chain mobility, polymer chain conformation, degree of polymer chain ordering or crystallinity can all vary significantly and continuously from the interface with the reinforcement into the bulk of the matrix.

This large amount of reinforcement surface area means that a relatively small amount of nanoscale reinforcement can have an observable effect on the macroscale properties of the composite. For example, adding carbon nanotubes improves the electrical and thermal conductivity. Other kinds of nanoparticulates may result in enhanced optical properties, dielectric properties, heat resistance or mechanical properties such as stiffness, strength and resistance to wear and damage. In general, the nanoreinforcement is dispersed into the matrix during processing. The percentage by weight (called mass fraction) of the nanoparticulates introduced can remain very low (on the order of 0.5 to 5 %) due to the low filler percolation threshold, especially for the most commonly used nonspherical, high aspect ratio fillers (*e. g.*, nanometer-thin platelets, such as clays, or nanometer-diameter cylinders, such as carbon nanotubes).

(1) Ceramic-matrix nanocomposites

In this group of composites the main part of the volume is occupied by a ceramic, *i. e.*, a chemical compound from the group of oxides, nitrides, borides, silicides, *etc*. In most cases, ceramic-matrix nanocomposites encompass a metal as the second component. Ideally both components, the metallic one and the ceramic one, are finely dispersed in each other in order to elicit the particular nanoscopic properties. Nanocomposite from these combinations was demonstrated in improving their optical, electrical and magnetic properties as well as tribological, corrosion-resistance and other protective properties.

The binary phase diagram of the mixture should be considered in designing ceramic-metal nanocomposites and measures have to be taken to avoid a chemical reaction between both components. The last point mainly is of importance for the metallic component: that may easily react with the ceramic and thereby loose its metallic character. This is not an easily obeyed constraint, because the preparation of the ceramic component generally requires high process temperatures. The safest measure thus is to carefully choose immiscible metal and ceramic phases. A good example for such a combination is represented by the ceramic-metal composite of TiO_2 and Cu, the mixtures of which were found immiscible over large areas in the Gibbs' triangle of Cu-O-Ti.

The concept of ceramic-matrix nanocomposites was also applied to thin films that are solid layers of a few nm to some tens of μm thickness deposited upon an underlying substrate and that play an important role in the functionalization of technical surfaces. Gas now sputtering by the hollow cathode technique turned out as a rather effective technique for the preparation of nanocomposite layers. The process operates as a vacuum-based deposition technique and is associated with high deposition rates up to some um/s and the growth of nanoparticles in the gas phase. Nanocomposite layers in the ceramics range of composition were prepared from TiO_2 and Cu by the hollow cathode technique that showed a high mechanical hardness, small coefficients of friction and a high resistance to corrosion.

(2) Metal-matrix nanocomposites

Another kind of nanocomposite is the energetic nanocomposite, generally as a hybrid sol-gel with a silica base, which, when combined with metal oxides and nanoscale aluminium powder, can form superthermite materials.

(3) Polymer-matrix nanocomposites

In the simplest case, appropriately adding nanoparticulates to a polymer matrix can enhance its performance, often in very dramatic degree, by simply capitalizing on the nature and properties of the nanoscale filler (these materials are better described by the term nanofilled polymer composites). This strategy is particularly effective in yielding high performance composites, when good dispersion of the filler is achieved and the properties of the nanoscale filler are substantially different or better than those of the matrix, for example, reinforcing a polymer matrix by much stiffer nanoparticles of ceramics, clays or carbon nanotubes. Alternatively, the enhanced properties of high performance nanocomposites may be mainly due to the high aspect ratio and/or the high surface area of the fillers, since nanoparticulates have extremely high surface area to volume ratios when good dispersion is achieved.

Nanoscale dispersion of filler or controlled nanostructures in the composite can introduce new physical properties and novel behaviours that are absent in the unfilled matrices, effectively changing the nature of the original matrix (such composite materials can be better described by the term genuine nanocomposites or hybrids). Some examples of such new properties are fire resistance or flame retardancy and accelerated biodegradability.

Chapter 15 Nanocomposite

15.2 Organic-Inorganic Nanocomposites

Organic-inorganic hybrid materials do not represent only a creative alternative to design new materials and compounds for academic research, but their improved or unusual features allow the development of innovative industrial applications. Nowadays, most of the hybrid materials that have already entered the market are synthesized and processed by using conventional soft chemistry based routes developed in the eighties. These processes are based on:

(a) The copolymerisation of functional organosilanes, macromonomers and metal alkoxides.

(b) The encapsulation of organic components within sol-gel derived silica or metallic oxides.

(c) The organic fictionalisation of nanofillers, nanoclays or other compounds with lamellar structures, *etc*.

The chemical strategies (self-assembly, nanobuilding block approaches, hybrid MOF (Metal organic frameworks), integrative synthesis, coupled processes, bioinspired strategies, *etc*.) offered nowadays by academic research allow, through an intelligent tuned coding, the development of a new vectorial chemistry, able to direct the assembling of a large variety of structurally well defined nanoobjects into complex hybrid architectures hierarchically organized in terms of structure and functions.

Independently of the types or applications, as well as the nature of the interface between organic and inorganic components, a second important feature in the tailoring of hybrid networks concerns the chemical pathways that are used to design a given hybrid material. General strategies for the synthesis of sol-gel derived hybrid materials have been already discussed in details in several reviews.

(1) Path A

Path A corresponds to very convenient soft chemistry based routes including conventional sol-gel chemistry, the use of specific bridged and polyfunctional precursors and hydrothermal synthesis.

Route A1: Via conventional sol-gel pathways amorphous hybrid networks are obtained through hydrolysis of organically modified metal alkoxides (vide infra section III) or metal halides condensed with or without simple metallic alkoxides. The solvent may or may not contain a specific organic molecule, a biocomponent or polyfunctional polymers that can be crosslinkable or that can interact or be trapped within the inorganic components through a large set of fuzzy interactions (H-bonds, π-π interactions, van der Waals②). These strategies are simple, low cost and yield amorphous nanocomposite hybrid materials. These materials, exhibiting infinite microstructures, call be transparent and easily shaped as films or bulks. They are generally poly disperse in size and locally heterogeneous in chemical composition. However, they are cheap, very versatile, present many interesting properties and consequently they give rise to many commercial products shaped as films, powders or monoliths. These commercial products and their field of application will be discussed in section III-2. Better academic understanding and control of the local and semilocal structure of the hybrid materials and their degree of organization are important issues, especially if in the future tailored properties are sought.

Route A2: The use of bridged precursors such as silsesquioxanes X_3Si-R'-SiX_3 (R' is an organic spacer, X = Cl, Br, OR) allow the formation of homogeneous molecular hybrid organic-inorganic materials which have a better degree of local organisation. In recent work, the organic spacer has been complemented by using two terminal functional groups (urea type). The combination within the organic bridging component of aromatic or alkyl groups and urea groups allows better self-assembly through the capability of the organic moieties to establish both strong hydrogen bond networks and efficient packing via π-π or hydrophobic interactions.

Route A3: Hydrothermal synthesis in polar solvents (water, formamide, *etc.*) in the presence of organic templates had given rise to numerous zeolites with an extensive number of applications in the domain of adsorbents or catalysts. More recently a new generation of crystalline microporous hybrid solids has been discovered by several groups. These hybrid materials exhibit very high surface areas (from 1 000 to 4 500 m^2/g) and present hydrogen uptakes of about 3.8 *wt* % at 77 K. Moreover, some of these new hybrids can also present magnetic or electronic properties. These hybrids MOF are very promising for catalytic and gas adsorption based applications.

(2) Path B

Path B corresponds to the assembling (route B1) or the dispersion (route B2) of well-defined nanobuilding blocks (NBB) which consists of perfectly calibrated preformed objects that keep their integrity in the final material. This is a suitable method to reach a better definition of the inorganic component. These NBB can be clusters, organically pre- or post-functionalized nanoparticles (metallic oxides; metals, chalcogenides, *etc.*), nano-core-shells or layered compounds (clays, layered double hydroxides, lamellar phosphates, oxides or chalcogenides) able to intercalate organic components. These NBB can be capped with polymerizable ligands or connected through organic spacers, like telechelic molecules or polymers, or functional dendrimers. The use of highly pre-condensed species presents several advantages: they exhibit a lower reactivity towards hydrolysis or attack of nucleophilic moieties than metal alkoxides; the nanobuilding components are nanometric, monodispersed, and with better defined structures, which facilitates the characterization of the final materials.

The variety found in the nanobuilding blocks (nature, structure and functionality) and links allows one to build an amazing range of different architectures and organic-inorganic interfaces, associated with different assembling strategies. Moreover, the step-by-step preparation of these materials usually allows for high control over their semilocal structure. One important set of the NNB based hybrid materials that are already on the market are those resulting from the intercalation, swelling and exfoliation of nanoclays by organic polymers.

(3) Path C

In the last ten years, a new field has been explored, which corresponds to the organization or the texturation of growing inorganic or hybrid networks, templated growth by organic surfactants (Route C1). The success of this strategy is also clearly related to the ability that materials scientists have to control and tune hybrid interfaces. In this field, hybrid organic-inorganic phases are very interesting due to the versatility they demonstrate in the building of a whole continuous range of nanocomposites, from ordered dispersions of inorganic bricks in a hybrid matrix to highly controlled nanosegregation of organic polymers within inorganic matrices. In the latter case, one of the most striking examples is the synthesis of mesostructured hybrid networks. A recent strategy developed by several groups consists of the templated growth (with surfactants) of mesoporous hybrids by using bridged silsesquioxanes as precursors (Route C2). This approach yields a new class of periodically organized mesoporous hybrid silicas with organic functionality within the walls. These nanoporous materials present a high degree of order and their mesoporosity is available for further organic functionalisation through surface grafting reactions.

Route C3 corresponds to the combination of self-assembly and NBB approaches. Strategies combining the nanobuilding block approach with the use of organic templates that self-assemble and allow one to control the assembling step are also appearing. This combination between "nanobuilding block approach" and "templated assembling" will have paramount importance in exploring the theme of "synthesis with construction". Indeed, they exhibit a large variety of interfaces between the organic and the inorganic components (covalent bonding,

Chapter 15 Nanocomposite

complexation, electrostatic interactions, *etc*.). These NBB with tunable functionalities can, through molecular recognition processes, permit the development of a new vectorial chemistry.

(4) Path D

Path D is integrative synthesis. The strategies reported above mainly offer the controlled design and assembling of hybrid materials in the 1 to 500 Å range. Recently, micromolding methods have been developed, in which the use of controlled phase separation phenomena, emulsion droplets, latex beads, bacterial threads, colloidal templates or organogelators lead to controlling the shapes of complex objects in the microscale. The combination between these strategies and those above described along paths A, B and C allow the construction of hierarchically organized materials in terms of structure and functions. These synthesis procedures are inspired by those observed in natural systems for some hundreds of millions of years. Learning the "savoir faire[③]" of hybrid living systems and organisms from understanding their rules and transcription modes could enable us to design and build ever more challenging and sophisticated novel hybrid materials.

◆ Notes

① organic-inorganic hybrid materials 有机-无机杂化材料

② van der Waals 范德华力

③ savoir faire 随机应变之道,能力才干

◆ Vocabulary

abalone [ˌæbəˈləʊnɪ] *n*. [无脊椎]鲍鱼

alkoxide [ælˈkɒksaɪd] *n*. [有化]醇盐;酚盐

alkyl [ˈælkaɪl] *n*. 烷基,烃基 *adj*. 烷基的,烃基的

biodegradability [ˈbaɪəʊdɪˌɡreɪdəˈbɪlətɪ] *n*. [生物]生物降解能力

calibrate [ˈkælɪbreɪt] *vt*. 校正;调整;测定口径

catalytic [ˌkætəˈlɪtɪk] *n*. 催化剂;刺激因素 *adj*. 接触反应的;起催化作用的

chalcogenide [ˈkælkədʒənaɪd] *n*. 氧属化物;硫族化物

colloidal [kəˈlɒɪdəl] *adj*. 胶体的;胶质的;胶状的

complexation [kɒmplekˈseɪʃ(ə)n] *n*. 络合;络合作用(complex 的名词)

copolymerization [kəʊˌpɒləmaraɪˈzeɪʃən] *n*. 共聚作用

cosmetics [kɒzˈmetɪks] *n*. [化工]化妆品(cosmetic 的复数);装饰品 *v*. 用化妆品打扮(cosmetic 的三单形式)

dielectric [ˌdaɪɪˈlektrɪk] *n*. 电介质;绝缘体 *adj*. 非传导性的;诱电性的

dislocation [ˌdɪslə(ʊ)ˈkeɪʃ(ə)n] *n*. 转位;混乱;[医]脱白

Chapter 15 Nanocomposite

electrostatic [ɪˌlektrə(ʊ)'stætɪk] adj. 静电的;静电学的

elicit [ɪ'lɪsɪt] vt. 抽出,引出;引起

encapsulation [ɪnˌkæpsə'leɪʃən] n. 封装;包装

halide ['heɪlaɪd] n. 卤化物 adj. 卤化物的

hierarchically [ˌhaɪə'rækɪkəlɪ] adv. 分层次,分等级地

latex ['leɪteks] n. 乳胶;乳液

mesoporosity [miːsə'pɔːrɒsɪtɪ] n. 中孔

mesoporous [miːsə'pɔːrəs] n. 介孔;孔直径在 2 到 50 纳米之间;介孔材料

moiety ['mɒɪɪtɪ] n. [化学]一部分;一半

monodispersed [mɒnəʊdɪs'pɜːst] adj. 单分散的(monodisperse 的变形)

monolith ['mɒn(ə)lɪθ] n. 整块石料;庞然大物

organosilane [ɔːɡənəʊ'zaɪlən] n. [有化]有机硅烷

platelet ['pleɪtlɪt] n. [组织]血小板;薄片

poly ['pɒlɪ] n. polytechnic 的缩写)工艺专科学校;科技学校;工业学校(或大学) adj. (polyester 的缩写)聚酯的,涤纶的

polyfunctional [ˌpɒlɪ'fʌŋkʃənəl] adj. 多官能的;多重的

retardancy [ri'tɑːdənsɪ] n. 阻滞性;阻止性

silsesquioxane [sɪlseskwɑ'ɪɒkseɪn] n. 倍半硅氧烷

superparamagnetism [ˈsjuːpəˌpærəˈmæɡnɪtɪzəm] n. [电磁]超顺磁性

surfactant [sə'fækt(ə)nt] n. 表面活性剂 adj. 表面活性剂的

transcription [træn'skrɪpʃ(ə)n] n. 抄写;抄本;誊写

triangle ['traɪæŋɡ(ə)l] n. 三角(形);三角关系;三角形之物;三人一组

tribological [ˌtraɪbəʊ'lɒdʒɪkəl] adj. 摩擦学的

urea [jʊ'riːə] n. [肥料]尿素

vectorial [vek'tɔːrɪəl] adj. [数]向量的;带菌体的

versatility [ˌvɜːsə'tɪlətiː] n. 多功能性;多才多艺;用途广泛

zeolite ['ziːəlaɪt] n. 沸石

Exercises

1. Translate the following Chinese phrases into English

(1) 催化性能 (2) 质量分数 (3) 化合物

(4) 高沉积速率 (5) 高比表面 (6) 金属氧化物

(7) 层状结构 (8) 矢量化学 (9) 路径 A

2. Translate the following English phrases into Chinese

(1) magnetic materials (2) dielectric property

(3) ceramic-matrix nanocomposites (4) carbon nanotubes

(5) hybrid materials (6) metal alkoxides

(7) integrative synthesis (8) inorganic components

(9) hydrothermal synthesis (10) hydrophobic interaction

3. **Translate the following Chinese sentences into English**

(1) 化学合成既可以在液相也可以在气相中进行,但在固相中很难,因为物质在液相和气相中的扩散速率在数量级上大于其在固相中的速率。

(2) 纳米技术在聚合科学中的研究并不陌生,因为在纳米技术时代之前的研究已经涉及纳米级尺寸,只是没有像最近的纳米技术一样被专门提及。

(3) 尽管纳米复合材料的增强性是它的主要方面,但是纳米复合材料的许多其它性能和潜在的应用也很重要,包括阻抗性、阻燃性、电/电学性能、膜性能、聚合混合物兼容性。

(4) 生物和聚合科学的融合,通常涉及一些模仿生物系统的纳米级聚合混合物和复合系统的设计。

(5) 机械研磨方法实际上属于材料的固态加工过程,是"自上而下"方法的一个实例。

4. **Translate the following English sentences into Chinese**

(1) Superparamagnetic nanoparticles may further couple their physical properties to the matrix material, resulting in new functionalities, e. g., movement, change in shape or thermoresponsive behaviour.

(2) In addition, there are procedures to manufacture mesoporous SiO_2 materials with well-defined pores and large surface areas which have a great impact on catalytic and separation purposes.

(3) A broad classification divides methods into either those which build from the bottom up, atom by atom, or those which construct from the top down using processes that involve the removal or reformation of atoms to create the desired structure.

(4) Sol-gel methods involve a set of chemical reactions which irreversibly convert a homogeneous solution of molecular reaction precursors (a sol) into an infinite molecular weight three-dimensional polymer (a gel) forming an elastic solid.

(5) Generally, molecular self-assembly is the spontaneous organization of relatively rigid molecules into structurally well-defined aggregates via weak, reversible interactions such as hydrogen bonds, ionic bonds and van der Waals bonds.

5. **Translate the following Chinese essay into English**

CVD 是一种化学气相反应过程,该过程是利用易挥发性金属的蒸发,并进行化学反应以形成理想的化合物,在气相中迅速冷却生成纳米粒子。CVD 法生成的纳米粒子具有纯度高、均匀性好、粒径小、分布窄、分散性好以及化学反应活性高等许多优点。

6. **Translate the following English essay into Chinese**

Nearly all of the principles of green chemistry can be readily applied to the design of nanoscale products. Many preparations of the building blocks in nanotechnology involve hazardous chemicals, low material conversions, high energy requirements and generation of hazardous substances. Thus, there are multiple opportunities to develop greener processes for the manufacture of these materials.

Green chemistry is the utilization of a set of principles that reduces or eliminates generation of hazardous substances in the design, manufacture, and application of chemical products. Green nanoscience and nanotechnology involve the application of green chemistry principles to design nanoscale products. A growing number of applications in nanoscience/nanotechnology are being developed to promise environment-friendly including new catalysts for environmental remediation, cheap and efficient photovoltaic technology, thermoelectric materials for cooling without refrigerants and nanocomposites for vehicles. Nanoscale sensors can also offer faster response times and lower detection limits, making on-site, real-time detection possible.

扫一扫,查看更多资料

Part IV
Measurement and Analysis of Materials

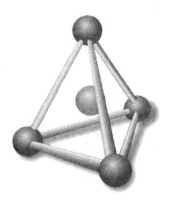

Chapter 16　Analytical Techniques of Materials

16.1　Introduction

For better understanding and utilizing various materials, it is very important to characterize them so as to find out the relationship between the properties and microstructures. There are many analytical techniques in the field of materials science. Some common methods will be discussed in this chapter.

16.2　X-Ray Diffraction

X-ray diffraction (XRD) is a rapid analytical technique primarily used for phase identification of a crystalline material and can provide information on unit cell dimensions. The analyzed material is finely ground, homogenized, and average bulk composition is determined.

Max von Laue, in 1912, discovered that crystalline substances act as three-dimensional diffraction gratings for X-ray wavelengths similar to the spacing of planes in a crystal lattice. XRD is now a common technique for the study of crystal structures and atomic spacing.

XRD is based on constructive interference of monochromatic X-rays and a crystalline sample. These X-rays are generated by a cathode ray tube, filtered to produce monochromatic radiation, collimated to concentrate, and directed toward the sample. The interaction of the incident rays with the sample produces constructive interference (and a diffracted ray) when conditions satisfy Bragg's law[①] (Figure 16.1).

$$n\lambda = 2d\sin\theta \tag{16-1}$$

Chapter 16 Analytical Techniques of Materials

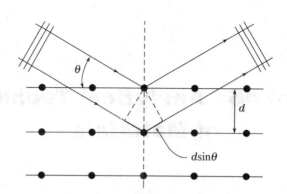

Figure 16.1 Schematic diagram of Bragg's law in XRD

This law relates the wavelength of electromagnetic radiation to the diffraction angle and the lattice spacing in a crystalline sample. These diffracted X-rays are then detected, processed and counted. By scanning the sample through a range of 2θ angles, all possible diffraction directions of the lattice should be attained due to the random orientation of the powdered material. Conversion of the diffraction peaks to d-spacings allows identification of the mineral because each mineral has a set of unique d-spacings. Typically, this is achieved by comparison of d-spacings with standard reference patterns.

The incoming beam (coming from upper left) causes each scatterer to reradiate a small portion of its intensity as a spherical wave. If scatterers are arranged symmetrically with a separation d, these spherical waves will be in sync (add constructively) only in directions where their path-length difference $2d\sin\theta$ equals an integer multiple of the wavelength λ. In that case, part of the incoming beam is deflected by an angle 2θ, producing a reflection spot in the diffraction pattern.

All diffraction methods are based on generation of X-rays in an X-ray tube. These X-rays are directed at the sample, and the diffracted rays are collected. A key component of all diffraction is the angle between the incident and diffracted rays. Powder and single crystal diffraction vary in instrumentation beyond this.

X-ray diffractometers consist of three basic elements: an X-ray tube, a sample holder and an X-ray detector (see Figure 16.2).

X-rays are generated in a cathode ray tube by heating a filament to produce electrons, accelerating the electrons toward a target by applying a voltage and bombarding the target material with electrons. When electrons have sufficient energy to dislodge inner shell electrons of the target material, characteristic X-ray spectra are produced. These spectra consist of several components, the most common being K_α and K_β. K_α consists, in part, of K_{α_1} and $K_{\alpha 2}$. $K_{\alpha 1}$ has a slightly shorter wavelength and twice the intensity as $K_{\alpha 2}$. The specific wavelengths are

Chapter 16 Analytical Techniques of Materials

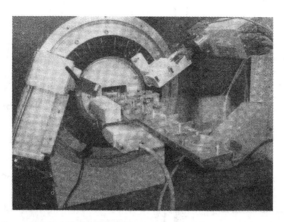

Figure 16.2 A bruker XRD D8 diffractometer with automatic sample changer (X-ray tube is on the left, the detector is on the right and the sample holder is in the middle)

characteristic of the target material (Cu, Fe, Mo and Cr). Filtering, by foils or crystal monochromoters, is required to produce monochrornatic X-rays needed for diffraction. K_{a_1} and K_{a_2} are sufficiently close in wavelength such that a weighted average of the two is used. Copper is the most common target material for single-crystal diffraction, with CuK_a radiation (1.541 8 Å). These X-rays are collimated and directed onto the sample. As the sample and detector are rotated, the intensity of the reflected X-rays is recorded. When the geometry of the incident X-rays impinging the sample satisfies the Bragg equation, constructive interference occurs and a peak in intensity occurs. A detector records and processes this X-ray signal and converts the signal to a count rate which is then output to a device such as a printer or computer monitor (see Figure 16.3).

The geometry of an X-ray diffractometer is such that the sample rotates in the path of the collimated X-ray beam at an angle θ while the X-ray detector is mounted on an arm to collect the diffracted X-rays and rotates at an angle of 2θ. The instrument used to maintain the angle and rotate the sample is termed a goniometer. For typical powder patterns, data is collected at 2θ from $-5°$ to $70°$, angles that are preset in the X-ray scan.

X-ray powder diffraction is most widely used for the identification of unknown crystalline materials (e.g., minerals, inorganic compounds). Determination of unknown solids is critical to studies in geology, environmental science, material science, engineering and biology.

Applications include:

(1) Characterization of crystalline materials.

(2) Determination of unit cell dimensions.

Chapter 16　Analytical Techniques of Materials

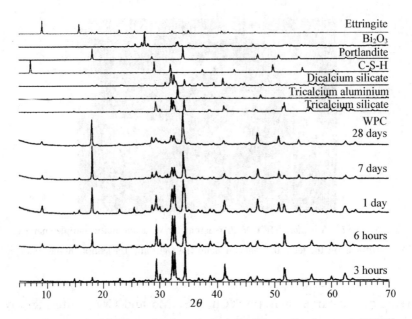

Figure 16.3　Some typical XRD results from hydrated white portland cement (WPC) pastes at various age, comparing with reference minerals

(It clearly shows how the contents of each mineral changes during the hydration, indicating the formation of ettringite and portlandite on the consuming of tricalcium silicate, dicalcium silicate and tricalcium aluminum)

(3) Measurement of sample purity.

With specialized techniques, XRD can be used to:

(1) Determine crystal structures using Rietveld refinement.

(2) Determine of modal amounts of minerals (quantitative analysis).

(3) Characterize thin films samples by:

① Determining lattice mismatch between film and substrate and to inferring stress and strain.

② Determining dislocation density and quality of the film by rocking curve measurements.

③ Measuring superlattices in multilayered epitaxial structures.

④ Determining the thickness, roughness and density of the film using glancing incidence X-ray reflectivity measurements.

(4) Make textural measurements, such as the orientation of grains, in a polycrystalline sample.

Chapter 16 Analytical Techniques of Materials

16.3 Optical Metallography

Optical metallography techniques are used to study the features and internal makeup of materials at the micrometer level (magnification level of around 2 000×). Qualitative and quantitative information pertaining to grains size, grain boundary, existence of various phases, internal damage, and some defects may be extracted using optical metallography techniques. In this technique, the surface of a small sample of a material such as a metal or a ceramic is first prepared through a detailed and rather lengthy procedure. The preparation process includes numerous surface grinding stages (usually four) that remove large scratches and thin plastically deformed layers from the surface of the specimen. The grinding stage is followed with a number of polishing stages (usually four) that remove fine scratches formed during the grinding stage. The quality of the surface is extremely important in the outcome of the process, and generally speaking, a smooth, mirror-like surface without scratches must be produced at the end of the polishing stage. These steps are necessary to minimize topographic contrast. The polished surface is then exposed to chemical etchants. The choice of the etchant and the etching time (the time interval in which the sample will remainin contact with the etchant) are two critical factors that depend on the specific material under study. The atoms at the grain boundary will be attacked at a much more rapid rate by the etchant than those atoms inside the grain. This is because the atoms at the grain boundary possess a higher state of energy because of the less efficient packing. As a result, the etchant produces tiny grooves along the boundaries of the grains. The prepared sample is then examined using a metallurgical microscope (inverted microscope) based on visible incident light. A schematic representation of the metallurgical microscope is given in Figure 16.4. When exposed to incident light in an optical microscope, these grooves do not reflect the light as intensely as the remainder of the grain material (Figure 16.5). Because of the reduced light reflection, the tiny grooves appear as dark lines to the observer, thus revealing the grain boundaries (Figure 16.6). Additionally, impurities, other existing phases, and internal defects also react differently to the etchant and reveal themselves in photomicrographs taken from the sample surface. Overall, this technique provides a great deal of qualitative information about the material.

Chapter 16 Analytical Techniques of Materials

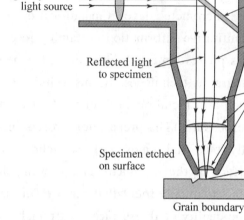

Figure 16.4 Schematic diagram illustrating how light is reflected from the surface of a polished and etched metal. The irregular surface of the etched-out grain boundary does not reflect light

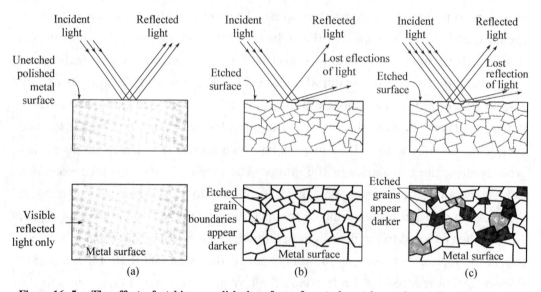

Figure 16.5 The effect of etching a polished surface of a steel metal sample on the microstructure observed in the optical microscope. (a) In the as-polished condition, no microstructural features are observed. (b) After etching a very low-carbon steel, only grain boundaries are chemically attacked severely, and so they appear as dark lines in the optical microstructure. (c) After etching a medium-carbon steel polished sample, dark (pearlite) and light (ferrite) regions are observed in the microstructure. The darker pearlite regions have been more severely attacked by the etchant and thus do not reflect much light

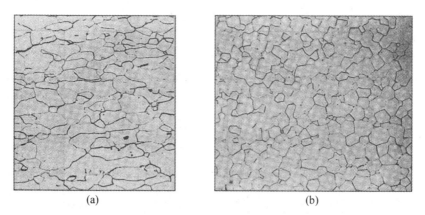

Figure 16.6 Grain boundaries on the surface of polished and etched samples as revealed in the optical microscope. (a) Low-carbon steel (magnification 100×). (b) Magnesium oxide (magnification 225×)

In addition to the qualitative information that is extracted from the photomicrographs, some limited quantitative information may also be extracted. Grain size and average grain diameter of the material may be determined using the photomicrographs obtained by this technique.

The grain size of polycrystalline metals is important since the amount of grain boundary surface has a significant effecton many properties of metals, especially strength. At lower temperatures (less than about one-half of their melting temperature), grain boundaries strengthen metals by restricting dislocation movement under stress. At elevated temperatures, grain boundary sliding may occur, and grain boundaries can become regions of weakness in polycrystalline metals.

One method of measuring grain size is the American Society for Testing Materials (ASTM) method, in which the grain-size number n is defined by

$$N = 2^{n-1} \tag{16-2}$$

where N is the number of grains per square inch on a polished and etched material surface at a magnification of $100\times$ and n is an integer referred to as the ASTM grain size number. Grain size numbers with the nominal number of grains per square inch at $100\times$ and grains per square millimeter at $1\times$ are listed in Table 16.1. Figure 16.7 shows some examples of nominal grain sizes for low-carbon sheet steel samples. Generally speaking, a material may be classified as coarse-grained when $n<3$; medium-grained, $4<n<6$; fine-grained, $7<n<9$ and ultrafine-grained, $n>10$.

Chapter 16 Analytical Techniques of Materials

Table 16.1 ASTM grain sizes

Grainsize no.	Nominal number of grains	
	Per sq mm at 1×	Per sq in. at 100×
1	15.5	1.0
2	31.0	2.0
3	62.0	4.0
4	124	8.0
5	248	16.0
6	496	32.0
7	992	64.0
8	1 980	128
9	3 970	256
10	7 940	512

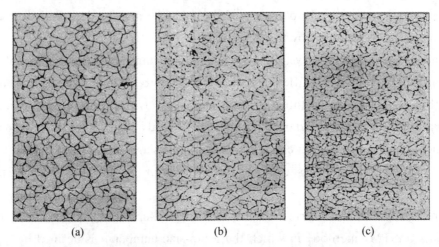

Figure 16.7 Several ASTM grain sizes of low-carbon sheet steels: (a) no. 7, (b) no. 8 and (c) no. 9. (Etch: nital; magnification 100×)

A more direct approach of assessing the grain size of a material would be to determine the actual average grain diameter. This offers clear advantages to the ASTM grain-size number that in reality does not offer any direct information about the actual size of the grain. In this approach, once a photomicrograph is prepared at a specific magnification, a random line of known length is drawn on the photomicrograph. The number of grains intersected by this line is then determined, and the ratio of the number of grains to the actual length of the line is determined, n_L. The average grain diameter d is determined using the equation:

$$d = \frac{C}{n_L M} \qquad (16-3)$$

where C is a constant ($C = 1.5$ for typical microstructures) and M is the magnification at which the photomicrograph is taken.

16.4 Scanning Electron Microscopy

Scanning electron microscope is an important tool in materials science and engineering; it is used for microscopic feature measurement, fracture characterization, microstructure studies, thin coating evaluations, surface contamination examination and failure analysis of materials. As opposed to optical microscopy where the sample's surface is exposed to incident visible light, the scanning electron microscopy (SEM) impinges a beam of electrons in a pinpointed spot on the surface of a target specimen and collects and displays the electronic signals given off by the target material. Figure 16.8 is a schematic illustration of the principles of operation of an SEM. Basically, an electron gun produces an electron

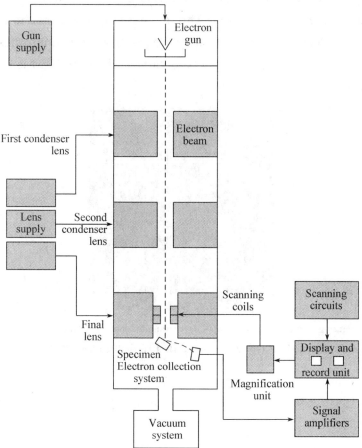

Figure 16.8 Schematic diagram of the basic design of a scanning electron

Chapter 16　Analytical Techniques of Materials

beam in an evacuated column that is focused and directed so that it impinges on a small spot on the target. Scanning coils allow the beam to scan a small area of the surface of the sample. Low-angle backscattered electrons interact with the protuberances of the surface and generate secondary backscattered electrons to produce an electronic signal, which in turn produces an image having a depth of field of up to about 300 times that of the optical microscope (about 10 μm at 10 000 diameters magnification). The resolution of many SEM instruments is about 5 nm, with a wide range of magnification (about 15 to 100 000×).

The SEM is particularly useful in materials analysis for the examination of fractured surfaces of metals. Figure 16.9 shows an SEM fractograph of an intergranular corrosion fracture. Notice how clearly the metal grain surfaces are delineated and the depth of perception. SEM fractographs are used to determine whether a fractured surface is intergranular (along the grain boundary) transgranular (across the grain), or a mixture of both. The samples to be analyzed using standard SEM are often coated with gold or other heavy metals to achieve better resolution and signal quality. This is especially important if the sample is made of a nonconducting material. Qualitative and quantitative information relating to the makeup of the sample may also be obtained when the SEM is equipped with an X-ray spectrometer.

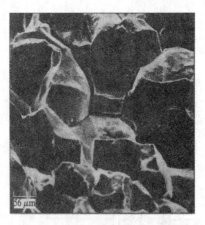

Figure 16.9　**Scanning electron fractograph of intergranular corrosion fracture near a circumferential weld in a thick-wall tube made of type 304 stainless steel (magnification 180×)**

Chapter 16 Analytical Techniques of Materials

16.5 Transmission Electron Microscopy

Transmission electron microscopy (TEM) in Figure 16.10 is an important technique for studying defects and precipitates (secondary phases) in materials. Much of what is known about defects would be speculative theory and would have never been verified without the use of TEM, which resolves features in the nanometer range.

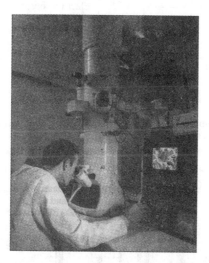

Figure 16.10　A man looking into an electron microscope

Defects such as dislocations can be observed on the image screen of a TEM. Unlike optical microscopy and SEM techniques where sample preparation is rather basic and easy to achieve, sample preparation for TEM analysis is complex and requires highly specialized instruments. Specimens to be analyzed using a TEM must have a thickness of several hundred nanometers or less depending on the operating voltage of the instrument. A properly prepared specimen is not only thin but also has fiat parallel surfaces. To achieve this, a thin section (3 to 0.5 mm) is cut out of the bulk material using techniques such as electric-discharge machining (used for conducting samples) and a rotating wire saw, among others. The specimen is then reduced to 50 μm thickness while keeping the faces parallel using machine milling or lapping processes with fine abrasives. Other more advanced techniques such as electropolishing and ion-beam thinning are used to thin a sample to its final thickness.

In the TEM, an electron beam is produced by a heated tungsten filament at the top of an evacuated column and is accelerated down the column by high voltage (usually from 100 to 300 kV). Electromagnetic coils are used to condense the electron beam, which is then passed through the thin specimen placed on the specimen stage. As the electrons pass through the specimen, some are absorbed and some are scattered so that they change direction. It is now clear that the sample thickness is critical: A thick sample will not allow the passage of electrons due to excessive absorption and diffraction. Differences in crystal atomic arrangements will cause electron scattering. After the electron beam has passed through the specimen, it is focused with the objective coil (magnetic lens) and then enlarged and projected on a fluorescent screen (Figure 16.11). An image can be formed either by

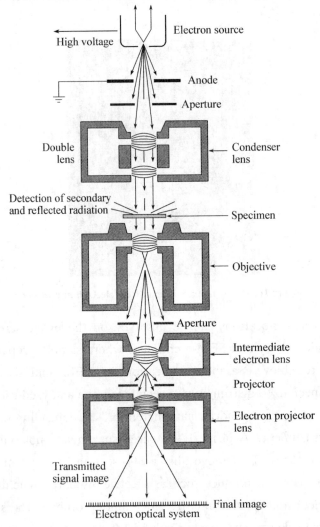

Figure 16.11 Schematic arrangement of electron lens system in a transmission electron microscope. All the lenses are enclosed in a column that is evacuated during operation. The path of the electron beam from the electron source to the final projected transmitted image is indicated by arrows. A specimen thin enough to allow an electron beam to be transmitted through it is placed between the condenser and objective lenses as indicated

collecting the direct electrons or the scattered electrons. The choice is made by inserting an aperture into the back focal plane of the objective lens. The aperture is maneuvered so that either the direct electrons or scattered electrons pass through it. If the direct beam is selected, the resultant image is called a bright-field image, and if the scattered electrons are selected, a dark-field image is produced.

In a bright-field mode, a region in a metal specimen that tends to scatter electrons to a higher degree will appear dark on the viewing screen. Thus, dislocations that have an irregular linear atomic arrangement will appear as dark lines on the electron microscope screen. A TEM image of the dislocation structure in a thin foil of iron deformed 14 percent at $-195\ ℃$ is shown in Figure 16.12.

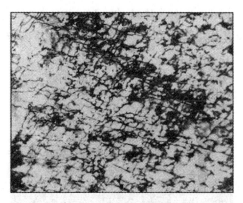

Figure 16.12 Dislocation structure in iron deformed 14 percent at $-195\ ℃$. The dislocations appear as dark lines because electrons have been scattered along the linear irregular atomic arrangements of the dislocations (Thin foil specimen; magnification 40 000×)

16.6 High-Resolution Transmission Electron Microscopy

Another important tool in the analysis of defects and crystal structure is the high-resolution transmission electron microscopy (HRTEM). The instrument has a resolution of about 0.1 nm, which allows viewing of the crystal structure and defects at the atomic level. To grasp what this degree of resolution may reveal about a structure, consider that the lattice constant of the silicon unit cell at approximately 0.543 nm is five times larger than the resolution offered by HRTEM. The basic concepts behind this technique are similar to those of the TEM. However, the sample must be significantly thinner on the order of 10 to 15 nm. In some situations, it is possible to view a two-dimensional projection of a crystal with the accompanying defects. To achieve this, the thin sample is tilted so that a low-index

Chapter 16 Analytical Techniques of Materials

direction in the plane is perpendicular to the direction of electron beam (atoms are exactly on top of each other relative to the beam). The diffraction pattern is representative of the periodic potential for the electrons in two dimensions. The interference of all the diffracted beams and the primary beam, when brought together again using the objective lens, provides an enlarged picture of the periodic potential. Figure 16.13 shows the HRTEM image of several dislocations (marked by "d") and some stacking faults (marked by arrows) in A&N thin film. In the figure, the periodic order of the atoms in the undisturbed regions is clearly observed (lower left). Dislocations create a wavy pattern in atomic structure. One can clearly see the disturbance in the atomic structure as a result of defects such as dislocations and stacking faults. It should be noted that due to limitations in the objective lens in HRTEM, accurate quantitative analysis of images is not easy to achieve and must be done with care.

Figure 16.13 HRTEM image of A & N atomic structure. The image shows two types of defects: (1) dislocations represented by arrows and the letter "d", and (2) stacking fault represented by two opposing arrows (top of the image)

16.7 Scanning Probe Microscopes and Atomic Resolution

Scanning tunneling microscope (STM) and atomic force microscope (AFM) are two of many recently developed tools that allow scientists to analyze and image materials at the atomic level. These instruments and others with similar capabilities are collectively classified as scanning probe microscopy (SPM). The SPM systems have the ability to magnify the features of a surface to the subnanometer scale, producing an atomic-scale topographic map of the surface. These instruments have important applications in many areas of science including but not limited to surface sciences where the arrangement of atoms and their bonding are important; metrology where surface roughness of materials needs to be analyzed; and nanotechnology where the position of individual atoms or molecules may be manipulated and new nanoscale phenomena may be investigated. It is appropriate to discuss these systems, how they function, the nature of information they provide, and their applications.

(1) Scanning tunneling microscope

IBM researchers G. Binnig and H. Rohrer developed the STM technique in the early 1980s and later received the Nobel Prize in physics in 1986 for their invention. In this technique, an extremely sharp tip (Figure 16.14), traditionally made of metals such as tungsten, nickel, platinum-iridium, or gold, and more recently out of carbon nanotubes, is used to probe the surface of a sample.

Figure 16.14 STM tip made of Pt-Ir alloy
(The tip is sharpened using chemical etching techniques)

The tip is first positioned a distance in the order of an atom diameter ($\cong$ 0.1 to 0.2 nm) from the surface of the sample. At such proximity, the electron clouds of the atoms in the tip of the probe interact with the electron clouds of the atoms on the surface of the sample. If at this point, a small voltage is applied across the tip and the surface, the electrons will "tunnel" the gap and, therefore, produce a small current that may be detected and monitored. Generally, the sample is analyzed under ultrahigh vacuum to avoid contamination and oxidation of its surface.

The produced current is extremely sensitive to the gap size between the tip and the surface. Small changes in the gap size produce an exponential increase in the detected current. As a result, small changes (less than 0.1 nm) in the position of the tip relative to the surface may be detected. The magnitude of current is measured when the tip is positioned directly above an atom (its electron cloud). This current is maintained at the same level as the tip moves over the atoms and valleys between the atoms (constant current mode) (Figure 16.15a). This is accomplished by adjusting the vertical position of the tip. The small movements required to adjust and maintain the current through the tip is then used to map the surface. The surface may also be mapped using a constant height mode in which the relative distance between the tip and the surface is maintained at a constant value and the changes in current are monitored (Figure 16.15b). The quality of the surface topography achieved by STM is striking as observed in the STM images of the surface of platinum (Figure 16.16).

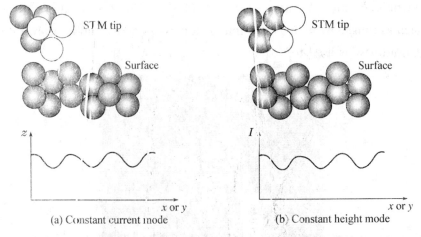

Figure 16.15 Schematics showing the STM modes of operation. (a) Adjust the z coordinate of the tip to maintain constant current (record z adjustments), (b) Adjust the current in the tip to maintain constant height (record I adjustments)

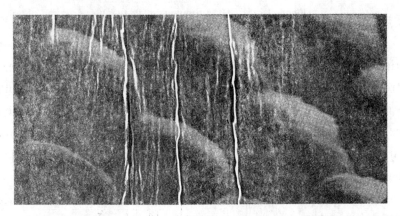

Figure 16.16 STM image of the surface of platinum showing outstanding atomic resolution

Clearly, what is extremely important here is that the diameter of the tip should be of the order of a single atom to maintain atomic-scale resolution. The conventional metal tips can easily be worn and damaged during the scanning process, which results in poor image quality. More recently, carbon nanotubes of about one to tens of nanometer in diameter are being used as nanotips for STM and AFM applications because of their slender structure and strength. The STM is primarily used for topography purposes, and it offers no quantitative insight into the bonding nature and properties of the material. Because the apparatus's function is based on creating and monitoring small amounts of current, only those materials that can conduct electricity may be mapped, including metals and semiconductors. However, many materials of high interest to the research community, such as biological materials or polymers, are not conductive and, therefore, cannot be analyzed using this technique. For nonconducting materials, the AFM techniques are applied.

(2) Atomic force microscope

The AFM uses a similar approach as the STM in that it uses a tip to probe the surface. However, in this case, the tip is attached to a small cantilever beam. As the tip interacts with the surface of the sample, the forces (Van der Waals forces) acting on the tip deflect the cantilever beam. The interaction may be a short-range repulsive force (contact mode AFM) or a long-range attractive force (non-contact-mode AFM). The deflection of the beam is monitored using a laser and a photodetector set up as shown in Figure 16.17. The deflection is used to calculate the force acting on the tip. During scanning, the force will be maintained at a constant level (similar to the constant current mode in STM), and the displacement of the tip will be monitored. The surface topography is determined from these small displacements. Unlike STM, the AFM approach does not rely on a current tunneling through the tip and can therefore be applied to all materials even nonconductors. This is the main advantage of AFM over its predecessor, STM. There are currently numerous other AFM-based techniques available with various imaging modes, including magnetic and acoustic. AFM in various imaging modes is being used in areas such as DNA research, in situ monitoring of corrosion in material, in situ annealing of polymers, and polymer-coating technology. The fundamental understanding of important issues in the above areas has been significantly enhanced because of the application of such techniques.

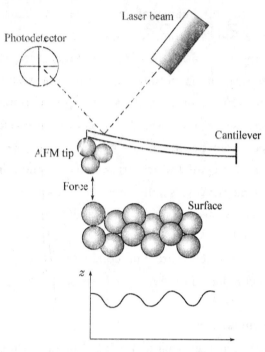

Figure 16.17 A schematic showing the basic AFM technique

Understanding the behavior of advanced materials at the atomic level drives the state of the art in high-resolution electron microscopy, which in turn provides an opportunity for the development of new materials. Electron and scanning probe microscopy techniques are and will be particularly important in nanotechnology and nanostructured materials.

16.8 Test of Mechanical Properties

The mechanical properties of a material are those properties that involve a reaction to an applied load. The mechanical properties of metals determine the range of usefulness of a material and establish the service life that can be expected. Mechanical properties are also used to help classify and identify material.

Most structural materials are anisotropic, which means that their material properties vary with orientation. The variation in properties can be due to directionality in the microstructure (texture) from forming or cold working operation, the controlled alignment of fiber reinforcement and a variety of other causes.

The mechanical properties of a material are not constants and often change as a function of temperature, rate of loading and other conditions. For example, temperatures below room temperature generally cause an increase in strength properties of metallic alloys; while ductility, fracture toughness and elongation usually decrease. Temperatures above room temperature usually cause a decrease in the strength properties of metallic alloys. Ductility may increase or decrease with increasing temperature depending on the same variables.

16.8.1 Loading Methods

The application of a force to an object is known as loading. Materials can be subjected to many different loading scenarios and a material's performance is dependent on the loading conditions. There are five fundamental loading conditions: tension, compression, bending, shear and torsion (Figure 16.18). Tension is the type of loading in which the two sections of material on either side of a plane tend to be pulled apart or elongated. Compression is the reverse of tensile loading and involves pressing the material together. Loading by bending involves applying a load in a manner that causes a material to curve and results in compressing the material on one side and stretching it on the other. Shear involves applying a load parallel to a plane which caused the material on one side of the plane to want to slide across the material on the other side of the plane. Torsion is the application of a force that causes twisting in a material.

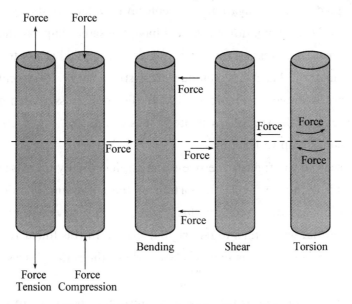

Figure 16.18　Fundamental loading conditions

If a material is subjected to a constant force, it is called static loading. If the loading of the material is not constant but instead fluctuates, it is called dynamic or cyclic loading. The way a material is loaded greatly affects its mechanical properties and largely determines how, or if, a component will fail; and whether it will show warning signs before failure actually occurs.

16.8.2 Tensile Strength

Tensile properties indicate how the material will react to forces being applied in tension. A tensile test is a fundamental mechanical test where a carefully prepared specimen is loaded in a very controlled manner while measuring the applied load and the elongation of the specimen over some distance. Tensile tests are used to determine the modulus of elasticity, elastic limit, elongation, proportional limit, reduction in area, tensile strength, yield point, yield strength and other tensile properties.

The main product of a tensile test is a load versus elongation curve which is then converted into a stress versus strain curve. The stress can be calculated by following equation.

$$\sigma = \frac{F}{A} \qquad (16-4)$$

Where F is the loading force, and A is the cross section area of the specimen.

Since both the engineering stress and the engineering strain are obtained by dividing the load and elongation by constant values (specimen geometry information), the load-elongation curve will have the same shape as the engineering stress-strain curve. The stress-strain curve relates the applied stress to the resulting strain and each material has its own unique stress-strain curve. A typical engineering stress-strain curve is shown in Figure 16.19. If the true stress, based on the actual cross-sectional area of the specimen, is used, it is found that the stress-strain curve increases continuously up to fracture.

As can be seen in the figure, the stress and strain initially increase with a linear relationship. This is the linear-elastic portion of the curve and it indicates that no plastic deformation has occurred. In this region of the curve, when the stress is reduced, the material will return to its original shape. In this linear region, the line obeys the relationship defined as Hooke's law[2] where the ratio of stress to strain is a constant.

The slope of the line in this region where stress is proportional to strain is called the modulus of elasticity or Young's modulus[3]. The modulus of elasticity (E) defines the properties of a material as it undergoes stress, deforms, and then returns to its original shape after the stress is removed. It is a measure of the stiffness of a

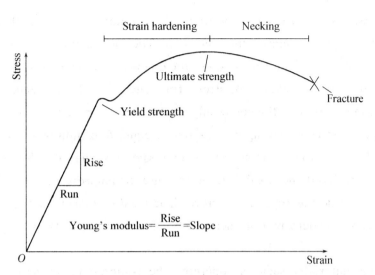

Figure 16.19 A Typical Engineering Stress-strain Curve

given material. To compute the modulus of elastic, simply divide the stress by the strain in the material. Since strain is unitless, the modulus will have the same units as the stress, such as MPa. The modulus of elasticity applies specifically to the situation of a component being stretched with a tensile force. This modulus is of interest when it is necessary to compute how much a rod or wire stretches under a tensile load.

With most materials there is a gradual transition from elastic to plastic behavior, and the exact point at which plastic deformation begins to occur is hard to determine. Therefore, various criteria for the initiation of yielding are used depending on the sensitivity of the strain measurements and the intended use of the data. For most engineering design and specification applications, the yield strength is used. The yield strength is defined as the stress required to produce a small, amount of plastic deformation. The offset yield strength is the stress corresponding to the intersection of the stress-strain curve and a line parallel to the elastic part of the curve offset by a specified strain (in the US the offset is typically 0.2 % for metals and 2 % for plastics).

16.8.3 Compressive Strength

In theory, the compression test is simply the opposite of the tension test with respect to the direction of loading. In compression testing the sample is squeezed while the load and the displacement are recorded. Compression tests result in mechanical properties that include the compressive yield stress, compressive ultimate stress and compressive modulus of elasticity.

Compressive yield stress is measured in a manner identical to that done for tensile yield strength. When testing metals, it is defined as the stress corresponding to 0.002 m/m plastic strain. For plastics, the compressive yield stress is measured at the point of permanent yield on the stress-strain curve. Moduli are generally greater in compression for most of the commonly used structural materials.

Ultimate compressive strength is the stress required to rupture a specimen. This value is much harder to determine for a compression test than it is for a tensile test since many materials do not exhibit rapid fracture in compression. Materials such as most plastics that do not rupture can have their results reported as the compressive strength at a specific deformation such as 1%, 5% or 10% of the sample's original height.

For some materials, such as concrete, the compressive strength is the most important material property that engineers use when designing and building a structure. Compressive strength is also commonly used to determine whether a concrete mixture meets the requirements of the job specifications.

16.8.4 Bend Strength

Bend strength, also known as modulus of rupture or flexural strength, a mechanical parameter for brittle material, is defined as a material's ability to resist deformation under load. The transverse bending test is most frequently employed, in which a rod specimen having either a circular or rectangular cross-section is bent until fracture using a three point flexural test technique. The flexural strength represents the highest stress experienced within the material at its moment of rupture. It is measured in terms of stress, here given the symbol σ.

When an object formed of a single material, like a wooden beam or a steel rod, is bent (Figure 16.20), it experiences a range of stresses across its depth. At the

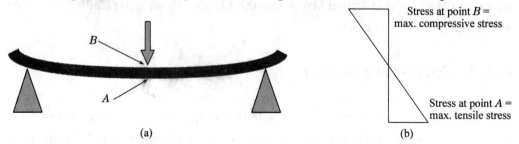

Figure 16.20 Diagram of beam under bending

(a) Beam of material under bending. Extreme fibers at B (compression) and A (tension), (b) Stress distribution across beam

edge of the object on the inside of the bend (concave face) the stress will be at its maximum compressive stress value. At the outside of the bend (convex face) the stress will be at its maximum tensile value. These inner and outer edges of the beam or rod are known as the "extreme fibers". Most materials fail under tensile stress before they fail under compressive stress, so the maximum tensile stress value that can be sustained before the beam or rod fails is its flexural strength.

The flexural strength would be the same as the tensile strength if the material were homogeneous. In fact, most materials have small or large defects in them which act to concentrate the stresses locally, effectively causing a localized weakness. When a material is bent only the extreme fibers are at the largest stress so, if those fibers are free from defects, the flexural strength will be controlled by the strength of those intact "ibers". However, if the same material was subjected to only tensile forces then all the fibers in the material are at the same stress and failure will initiate when the weakest fiber reaches its limiting tensile stress. Therefore it is common for flexural strengths to be higher than tensile strengths for the same material. Conversely, a homogeneous material with defects only on its surfaces (e. g., due to scratches) might have a higher tensile strength than flexural strength.

If we don't take into account defects of any kind, it is clear that the material will fail under a bending force which is smaller than the corresponding tensile force. Both of these forces will induce the same failure stress, whose value depends on the strength of the material.

The bend strength can be calculated by the following two equations. For a rectangular cross section specimen:

$$\sigma_f = \frac{3FL}{2bd^2} \tag{16-5}$$

For a circular cross section specimen:

$$\sigma_f = \frac{FL}{\pi R^3} \tag{16-6}$$

Where F—Load force, N.

L—Support span, mm.

b—Width of test beam, mm.

d—Depth of test beam, mm.

R—Radius of the beam, mm.

16.9 Test of Thermal and Electrical Properties

Thermal properties are the characteristics of a material that determine how it reacts when it is subjected to excessive heat, or heat fluctuations over time.

16.9.1 Thermal Conductivity

Thermal conductivity (λ) is the intrinsic property of a material which relates its ability to conduct heat. Heat transfer by conduction involves transfer of energy within a material without any motion of the material as a whole. Conduction takes place when a temperature gradient exists in a solid (or stationary fluid) medium. Conductive heat flow occurs in the direction of decreasing temperature because higher temperature equates to higher molecular energy or more molecular movement. Energy is transferred from the more energetic to the less energetic molecules when neighboring molecules collide.

Thermal conductivity is defined as the quantity of heat (Q) transmitted through a unit thickness (L) in a direction normal to a surface of unit area (A) due to a unit temperature gradient (ΔT) under steady state conditions and when the heat transfer is dependent only on the temperature gradient. In equation form this becomes the following:

$$\lambda = \frac{Q \cdot L}{A \cdot \Delta T} \qquad (16-7)$$

Thermal conductivity is important in material science, research, electronics, building insulation and related fields, especially where high operating temperatures are achieved. Several materials are shown in Figure 16.21. These should be considered approximate due to the uncertainties related to material definitions.

High energy generation rates within electronics or turbines require the use of materials with high thermal conductivity such as copper, aluminum and silver. On the other hand, materials with low thermal conductance, such as polystyrene and alumina, are use in building construction or in furnaces in an effort to slow the flow of heat, i.e., for insulation purposes.

The laser flash method is used to measure thermal diffusivity of a thin disc in the thickness direction. This method is based upon the measurement of the temperature rise at the rear face of the thin-disc specimen produced by a short energy pulse on the front face. With a reference sample specific heat can be

Chapter 16 Analytical Techniques of Materials

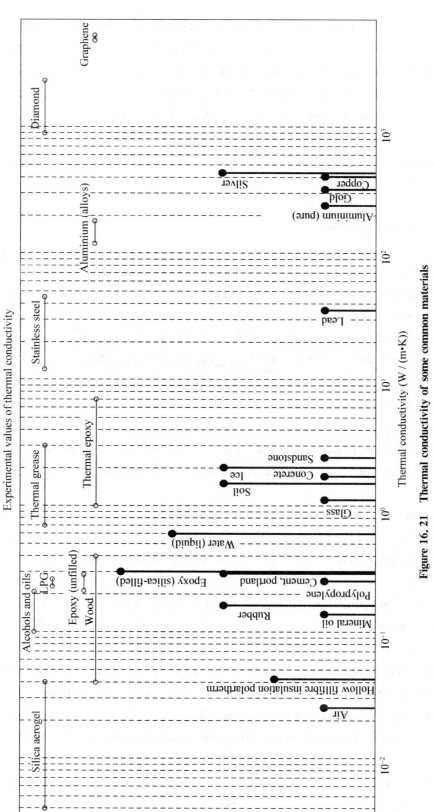

Figure 16.21 Thermal conductivity of some common materials

achieved and with known density the thermal conductivity results as follows.
$$\kappa(T) = \alpha(T) \cdot C_P(T) \cdot \rho(T) \tag{16-8}$$
Where κ—The thermal conductivity of the sample, W/(m·K).

α —The thermal diffusivity of the sample, m²/s.

C_P—The specific heat of the sample, J/(kg·K).

ρ—The density of the sample, kg/m³.

It is suitable for a multiplicity of different materials over a broad temperature range (-120 to 2 800 ℃).

16.9.2 Heat Capacity

Heat capacity or thermal capacity is the measurable physical quantity that specifies the amount of heat required to change the temperature of an object or body by a given amount. The SI unit of heat capacity is joule per kelvin, J/K.

Heat capacity is an extensive property of matter, meaning it is proportional to the size of the system. When expressing the same phenomenon as an intensive property, the heat capacity is divided by the amount of substance, mass or volume, so that the quantity is independent of the size or extent of the sample. The molar heat capacity is the heat capacity per mole of a pure substance and the specific heat capacity, often simply called specific heat, is the heat capacity per unit mass of a material. Occasionally, in engineering contexts, the volumetric heat capacity is used.

The heat capacity of most systems is not a constant. Rather, it depends on the state variables of the thermodynamic system under study. In particular it is dependent on temperature itself, as well as on the pressure and the volume of the system.

Different measurements of heat capacity can therefore be performed, most commonly either at constant pressure or at constant volume. The values thus measured are usually subscripted (by p and V, respectively) to indicate the definition. Gases and liquids are typically also measured at constant volume. Measurements under constant pressure produce larger values than those at constant volume because the constant pressure values also include heat energy that is used to do work to expand the substance against the constant pressure as its temperature increases.

There are several simple methods for measuring the specific heat capacities of both solids and liquids, one of which is electrical calorimeter.

Figure 16.22 shows possible arrangements for electrical calorimeters for a solid and a liquid specimen.

The material under investigation is heated by an electrical immersion heater and the input energy (H) and the rise in temperature that this produces are measured. If the mass of the specimen (solid or liquid) is m and its specific heat capacity c, then

$$Q = mc(T_1 - T_0) + q \qquad (16-9)$$

where T_0 and T_1 are the initial and final temperatures of the specimen and q is the heat loss. Using the cooling correction, the value of q may be found. This simple method can be used for liquids or solids, although in the case of a liquid allowance has to be made for the thermal capacity of the container, and the liquid should also be stirred to allow an even distribution of the heat energy throughout its volume. This is necessary since liquids are such poor conductors.

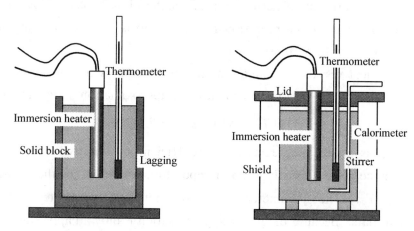

Figure 16.22 Calorimeter for measuring specific heat capacity of solid and liquid

16.9.3 Electrical Resistivity and Conductivity

Electrical resistivity (also known as resistivity, specific electrical resistance or volume resistivity) quantifies how strongly a given material opposes the flow of electric current. A low resistivity indicates a material that readily allows the movement of electric charge. Resistivity is commonly represented by the Greek letter ρ. The SI unit of electrical resistivity is the $\Omega \cdot m$ although other units like $\Omega \cdot cm$ are also in use. As an example, if $1\,m \times 1\,m \times 1\,m$ solid cube of material has sheet contacts on two opposite faces, and the resistance between these contacts is $1\,\Omega$, then the resistivity of the material is $1\,\Omega \cdot m$.

Electrical conductivity or specific conductance is the reciprocal of electrical resistivity, and measures a material's ability to conduct an electric current. It is commonly represented by the Greek letter σ, but κ (especially in electrical engineering) or γ are also occasionally used. Its SI unit is S/m.

The larger the cross-sectional area of the conductor, the more electrons per unit length is available to carry the current. As a result, the resistance is lower in larger cross-section conductors. The number of scattering events encountered by an electron passing through a material is proportional to the length of the conductor. The longer the conductor, therefore, the higher the resistance. Different materials also affect the resistance.

The electrical resistivity of a metallic conductor decreases gradually as temperature is lowered. In ordinary conductors, such as copper or silver, this decrease is limited by impurities and other defects. Even near absolute zero (0 K), a real sample of a normal conductor shows some resistance. In a superconductor, the resistance drops abruptly to zero when the material is cooled below its critical temperature. An electric current flowing in a loop of superconducting wire can persist indefinitely with no power source.

The electrical resistivity of most materials changes with temperature. If the temperature T does not vary too much, a linearapproximation is typically used:

$$\rho(T) = \rho_0[1 + \alpha(T - T_0)] \qquad (16-10)$$

Where α is called the temperature coefficient of resistivity, T_0 is a fixed reference temperature (usually room temperature), and ρ_0 is the resistivity at temperature T_0. The parameter α is an empirical parameter fitted from measurement data. Because the linear approximation is only an approximation, α is different for different reference temperatures. For this reason it is usual to specify the temperature that α was measured at with a suffix, such as α_{15}, and the relationship only holds in a range of temperatures around the reference. When the temperature varies over a large temperature range, the linear approximation is inadequate and a more detailed analysis and understanding should be used.

An electrical conductivity meter (EC meter) measures the electrical conductivity in a solution. Commonly used in hydroponics, aquaculture and freshwater systems to monitor the amount of nutrients, salts or impurities in the water.

The common laboratory conductivity meters employ a potentiometric method and four electrodes. Often, the electrodes are cylindrical and arranged concentrically. The electrodes are usually made of platinum metal. An alternating current is applied to the outer pair of the electrodes. The potential between the inner pair is measured. Conductivity could in principle be determined using the distance between the electrodes and their surface area using the Ohm's law[④] but generally, for accuracy, a calibration is employed using electrolytes of well-known conductivity.

Industrial conductivity probes often employ an inductive method, which has the advantage that the fluid does not wet the electrical parts of the sensor. Here, two inductively-coupled coils are used. One is the driving coil producing a magnetic field and it is supplied with accurately-known voltage. The other forms a secondary coil of a transformer. The liquid passing through a channel in the sensor forms one turn in the secondary winding of the transformer. The induced current is the output of the sensor.

Notes

① Bragg's Law 布拉格定律
② Hooke's law 胡克定律
③ Young's modulus 杨氏模量
④ Ohm's law 欧姆定律

Vocabulary

abrasive [ə'breɪsɪv] *adj*. 粗糙的;有研磨作用的;伤人感情的 *n*. 研磨料
acoustic [ə'kuːstɪk] *n*. 原声乐器;不用电传音的乐器 *adj*. 声学的;音响的;听觉的
backscatter ['bækskætə] *n*. [物]反向散射;背反射
bombard [bɒm'bɑːd] *n*. 射石炮 *vt*. 轰炸;炮击
calibration [kælɪ'breɪʃ(ə)n] *n*. 校准;刻度;标度
cantilever ['kæntɪliːvə] *n*. 悬臂
collimate ['kɒlɪmeɪt] *vt*. 校准;瞄准;使成平行
criteria [kraɪ'tɪərɪə] *n*. 标准,条件 (criterion 的复数)
diffraction [dɪ'frækʃn] *n*. (光,声等的)衍射,绕射
diffractometer [dɪfræk'tɒmɪtə] *n*. [光]衍射计
dislodge [dɪs'lɒdʒ] *vt*. 逐出,驱逐;使……移动;用力移动 *vi*. 驱逐,逐出;离开原位
etchant ['etʃ(ə)nt] *n*. 蚀刻剂
ettringite ['etrɪŋaɪt] *n*. [矿物]钙矾石;钙铝矾
fluctuate ['flʌktʃueɪt] *vi*. 波动;涨落;动摇 *vt*. 使波动;使动摇
fractograph ['fræktəʊgrɑːf] *n*. 断口组织;断口组织的显微镜照片
goniometer [ɡəʊnɪ'ɒmɪtə] *n*. 测角器,[仪]角度计
metallography [metə'lɒɡrəfɪ] *n*. [材]金相学
offset ['ɒfset] *n*. 抵消,补偿;平版印刷;支管 *vt*. 抵消;弥补;用平版印刷术印刷 *vi*. 装支管
pinpoint ['pɪnpɔɪnt] *n*. 针尖;精确位置;极小之物 *vt*. 查明;精确地找到;准确描述 *adj*. 精确的;详尽的
platinum ['plætɪnəm] *n*. [化学]铂;白金

Chapter 16　Analytical Techniques of Materials

portlandite ['pɔːtləndaɪt] n. [矿物]氢氧钙石
potentiometric [pəʊˌtenʃɪəˈmetrɪk] adj. 电势测定的,电位计的
precipitate [prɪˈsɪpɪteɪt] n. [化学]沉淀物 vt. 使沉淀;促成;猛抛;使陷入 adj. 突如其来的;猛地落下的;急促的 vi. [化学]沉淀
protuberance [prəˈtjuːb(ə)r(ə)ns] n. 突起,突出;结节
scatterer [ˈskætərə] n. [物]散射体;散射物质;扩散器
spectrometer [spekˈtrɒmɪtə] n. [光]分光仪
topographic [ˌtɒpəˈɡræfɪk] adj. 地形测量的;地志的;地形学上的
transgranular [trænsˈɡrænjʊlə] adj. 穿晶的;晶内的

◆ Exercises

1. Translate the following Chinese phrases into English

(1) 分析技术　　　　　(2) 相识别　　　　　(3) 路径长度差
(4) 单晶衍射　　　　　(5) 化学腐蚀剂　　　(6) 腐蚀面
(7) 透射电子显微镜　　(8) 电子束　　　　　(9) 样品厚度
(10) 扫描隧道显微镜　 (11) 扫描探针显微镜　(12) 超高真空
(13) 短程排斥力　　　 (14) 力学性能　　　　(15) 静载荷
(16) 载荷伸长曲线　　 (17) 线性关系　　　　(18) 弯曲强度
(19) 挠曲强度　　　　 (20) 热导率　　　　　(21) 热容

2. Translate the following English phrases into Chinese

(1) X-ray diffraction　　　　　　　(2) crystalline material
(3) diffraction pattern　　　　　　(4) optical metallography
(5) incident light　　　　　　　　(6) scanning electron microscopy
(7) ion-beam thinning　　　　　　 (8) bright-field image
(9) high-resolution transmission electron microscopy
(10) atomic force microscopy
(11) carbon nanotubes　　　　　　 (12) atomic-scale resolution
(13) long-range attractive force　(14) applied load
(15) tensile strength　　　　　　 (16) stress-strain curve
(17) compressive strength　　　　 (18) convex face
(19) depth of test beam　　　　　 (20) temperature gradient
(21) electrical resistivity

3. Translate the following Chinese sentences into English

(1) 因为电子穿过样本从而传递相关晶体结构、束缚态等大量信息,所以透射电子显微镜被广泛应用于材料科学、生物科学等领域。

(2) 自有人类文明以来,人们就一直为探索微观自然界和合成材料来认识和分析在不同类型相互作用条件下它们的性质和作用而不懈地努力。

(3) 一些材料在被电子束轰击时,能在可见光光谱、红外光谱或紫外光谱内发射出可见长波长光子。

(4) 许多物质或材料以化合物或混合物的形式存在。

(5) 热重分析法可以测量材料重量以及变化速率在受控条件下随时间以及温度的变化。

(6) 一些处于基态的分子在吸收高能量的光后跃迁到较高的能级上去,被称为"激发态"。

(7) 荧光显微镜是利用荧光现象而不是光学反射或光学吸附的一种光学设备或光学显微镜。

4. Translate the following English sentences into Chinese

(1) Since a single crystal shows a periodic lattice arrangement, material characteristics such as electrical conductivity, thermal expansivity, and hardness differ in each of the crystal directions.

(2) In the TEM apparatus, an objective aperture is set up in the back focal plane of the objective lens, and the images reflecting crystal information are obtained by adjusting the aperture to each focal position.

(3) The use of an electron beam for imaging purposes is only possible in a vacuum and with some special electron manipulation techniques for the beam to guarantee the direction of electron motion and focusing requirements.

(4) Melting is an endothermic transition because heat is required to break the ordered crystalline structure in the solid state for the sample to become a disordered structure in the liquid state.

(5) X-ray diffraction (XRD) is an important tool to investigate crystalline structures of materials.

(6) Variety of thermal transitions and thermodynamic interactions can be determined by means of DSC or ITC.

5. Translate the following Chinese essay into English

在日常生活中我们使用不同的材料——金属、陶瓷、高分子聚合物、半导体和复合材料。我们研究材料系统的功能、材料的加工和材料在工艺应用中的工程设计。我们应用化学、物理和生物学的基本原理来理解材料在相关的长度尺度下的结构和这样的结构所决定的性能。我们设计了科学的过程来制造材料以满足在特殊状况下的应用需求。可能该学科的最大的特征在鉴别和集中在加工、结构、性能和形貌间相互作用和相互依赖的关系上。材料制备和材料表征构成树料研究和精密仪器的重要方面,可以应用于过程、机制和材料的相互作用。

6. Translate the following English essay into Chinese

The scanning tunneling microscope (STM) was invented by Binnig and Rohrer in 1982, for which they were awarded the Nobel Prize in 1986. Later, the atomic force microscope (AFM) based on the principle of the STM was developed, for the research purpose concerned with surface structures of noconducting and conducting materials. STM and AFM are proved to be powerful tools for obtaining information of atomic morphology on a surface of material. In addition, by use of the nanometer manipulation, one has applied STM and AFM for manufacturing nanosized objects or devices.

扫一扫,查看更多资料

Part V
Application of Materials

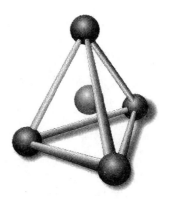

Part V

Application of Materials

Chapter 17 Application of Materials

17.1 Application of Metal Materials

17.1.1 Used as Structural Materials or Components

For modern engineering purposes, steel is the cheapest and most versatile of metals. Steels are available in thousands of types, ranging from hard to soft, can be magnetic or nonmagnetic, are heat-treatable and weldable, and have various resistances to heat, corrosion, impact and abrasion. The familiar products are steel reinforced concrete, stainless steels and tool steels.

The uses of some ferrous metals are listed according to their carbon content in Table 17.1.

Table 17.1 Use of ferrous metals by carbon content

Type	Carbon range / wt %	Typical uses
Carbon steels		
Low	0.05 to 0.30	For cold formability Wire, nails, rivets, screws Sheet stock for drawing Fenders, pots, pans, welding rods Bars, plates, structural shapes, shafting Forgings, carburized parts, key-stock
Medium	0.30 to 0.60	Free-machining steel Heat-treated parts that require moderate strength and high toughness such as bolts, shafting, axles, spline shafts Higher strength, heat-treated parts with moderate toughness such as lock washers, springs, band saw blades, ring gears, valve springs, snap rings

Chapter 17 Application of Materials

(to be continued)

Type	Carbon range / wt %	Typical uses
High	0.60 to 2.0	Chisels, center punches
		Music wire, mower blades, leaf springs
		Hay rake tines, leaf springs, knives, woodworking tools, files, reamers
		Ball bearings, punches, dies
Cast irons		
Gray	2.0 to 4.5	Machinable castings such as engine blocks, pipe, gears, lathe beds
White	2.0 to 3.5	Nonmachinable castings such as cast parts for wear resistance
Malleable castings	2.0 to 3.5	Produced from white cast iron. Machinable castings such as axle and differential housings, crankshafts, camshafts
Nodular iron (ductile iron)	2.0 to 4.5	Machinable castings such as pistons, cylinder blocks and heads, wrenches, forming dies

17.1.2 Metals in Electrical and Electronics Applications

Metals are very useful in electrical and electronics applications. For examples, high electrical conductivities make aluminum and copper good conductors. At very low temperatures, approaching absolute zero, copper becomes a superconductor. Magnetic resonance imaging (MRI) devices used in hospitals for diagnosing patients are a current example of a medical application of superconductivity.

Table 17.2 lists the names and typical uses of some of the more common copper alloys generally available.

Table 17.2 **Coppers and copper alloys**

Name	Applications
Coppers	
Electrolytic tough pitch copper (ETP)	Roofing, nails, rivets, cotter pins, kettles
Phosphorous deoxidized copper (DHP)	Piping, tubing, heat-transfer equipment, tanks
Free-cutting copper	Electrical connectors, motor and switch
	Parts, oxyacetylene torch tips
Oxygen-free (OF) copper	Electrical conductors, bus bars, tubing
Brasses	
Red brass	Trim, conduit and sockets, fire extinguishers
Cartridge brass	Radiator cores and tanks, reflectors, munitions

Chapter 17 Application of Materials

(to be continued)

Name	Applications
Yellow brass	Radiators, lamp fixtures, plumbing supplies
Medium-leaded brass	Engravings, gears, nuts, couplings
High-leaded brass	Nuts, gears, wheels, clock parts
Free-cutting brass	Gears and pinions
Muntz metal	Hardware, panels and sheets, forgings
Naval brass	Nuts and bolts, propeller shafts
Forged brass	Forgings, valve stems, plumbing supplies
Bronzes	
Aluminum bronze	Sheets, wire, tubing
Silicon bronze-grade A	Hydraulic lines, marine hardware
Silicon bronze-grade B	Hydraulic lines, marine hardware
Phosphor bronze-grade A	Bearing plates, clutch discs, bushings
Phosphor bronze-grade C	Springs, brushes, textile machine parts
Aluminum-silicon bronze	Forgings and extrusions
Phosphor bronze-grade D	Bars and plates, fittings
Commercial bronze	Munitions, jewelry, plaques and awards
Leaded commercial bronze	Screws, hardware
Cupro-nickel	Condensers, ferrules, tanks, valves, auto parts
Nickel-silver	Camera parts, plates, fixtures, tableware, zippers
Manganese bronze	Shafts, clutch parts, valve stems, forgings, welding rods
Architectural bronze	Trim, extrusions, hinges, auto parts

17.1.3 Metals in Aerospace

(1) Aluminum alloys

The majority of materials used in subsonic aircrafts are aluminum alloys, about $70\sim80$ wt % in commercial airplanes and $40\sim60$ wt % in military aircraft.

Aluminum lithium alloy is identified by its low density, high specific strength, high stiffness to weight ration, excellent low temperature performance, good corrosion resistance and excellent super-plastic formability. Using aluminum lithium alloy can provide a 15 % reduction in weight below aluminum alloy structures, and a $15\sim20$ % increase in stiffness. It is considered as an ideal structural material in aerospace industry. It is forecasted that aluminum alloys are the potential important structural materials in aircraft of future.

(2) Precious metals in aerospace

In the fields requiring materials with high temperature resistance, high reliability, high precision and long service life, such as aerospace, marine industries, only noble metals satisfy the requirement. For examples, rocket ignition alloys, aviation engine ignition contact, direction controlling and instrument materials in missiles, satellites, ships and aircrafts require accurate temperature measuring and shape-sensitive materials.

(3) Titanium alloys

Titanium alloys offer superior specific strength at high temperatures (over 590 ℃) and low temperatures (-253 ℃), which makes it a popular structural metal in ultra-high speed aircraft and accounts for its use on the space shuttle.

17.1.4 On the Road: Metals in Transportation

(1) Aluminum alloys

Due to the properties of high strength, high toughness, high modulus and good corrosion resistant, aluminum and its alloy are the light metals firstly and widely used in automobiles. They are also the most economical and competitive light metals for cars, from the viewpoint of cost and quality.

Audi A2 with a new lightweight structure was firstly manufactured by 100 *wt* % of aluminum in 1999 by Germany Audi Company.

(2) Magnesium alloys

Magnesium is a light silvery white metal. By alloying with other metals, such as aluminum, zinc and manganese, it will become a light alloy with high strength, high rigidity and good casting properties. Magnesium alloys have been widely used in modern automobiles.

By new processes, impurities of iron, copper, nickel and other elements are strictly limited. The resultant magnesium alloy has an improvement in corrosion resistance and a reduction at cost, which greatly promotes its applications in automobile. It is generally used on car seat frames, instruments, steering wheels, steering columns, engine cylinder covers, gearbox shell and clutch.

The advantages to use magnesium alloy parts are obviously. First, it is lighter than aluminum and steel and can reduce fuel consumption. Second, its specific strength is higher than that of aluminum alloy and steel. The third is that it has a good casting property and dimensional stability, good machinability, and low waste rate, thereby reducing production cost. The last is that it has a good damping coefficient to reduce noise and vibration, and to improve vehicle safety and comfortability. However, the cost of magnesium alloys is higher than that of aluminum alloys.

(3) Titanium

Titanium is used in almost all of current racing cars because of its lightweight and high strength. With titanium alloy valve, not only the weight of car can be reduced, the service life can be prolonged, but also the reliability of car is high, and fuel can be saved.

17.1.5 Metals and Medicine

(1) Titanium

Biomaterials are materials that are compatible with human and animal systems, allowing the material to be implanted or manipulated in people and animals. Titanium is an example of a biomaterial that serves as implants for joint replacements and dental reconstruction.

Titanium is used in the medical profession. It is lighter than stainless steel when used in medical instruments and is used in artificial replacement joints for two reasons: It is the most biocompatible metal known, and it has the stiffness similar to that of the human bone. Titanium is often used in conjunction with polyethylene and high-purity ceramics. Titanium forms a thin layer of oxide, which protects against further corrosion.

(2) Precious metals in biomedicine

Noble metals have excellent physical and chemical properties, such as high temperature oxidation resistance and corrosion resistance, good electrical conductivity, high catalytic activity, strong coordination ability. They are widely used in industry, and known as the modern "industrial vitamin".

Besides non-toxic and compatible to human body system, the devices implanted into human body also require good corrosion resistance, resistance to creep, and so on. The noble metals of platinum and its alloys, especially microprobe to explore nervous system and to repair damaged part, already succeed.

Gold drug is mainly used for treatment of rheumatoid arthritis. There are 27 kinds of drugs which are sold on world markets. Recently, silver containing drugs has a greater progress in the treatment of burns, caries prevention, breast cancer diagnosis, inhibition of HIV/AIDS transmission and laser surgery operations.

Chapter 17 Application of Materials

17.2 Application of Ceramics Materials

17.2.1 Refractories

We all know that metals and their alloys are used to make automobiles, machinery, planes, buildings and a myriad of other useful things possible. But metal production would not be possible without the use of a ceramic material called refractories. Refractories can withstand volatile and high-temperature conditions encountered in the processing of metals. Refractory ceramics are enabling materials for other industries as well. The chemical, petroleum, energy conversion, glass and other ceramic industries all rely upon refractory materials.

17.2.2 Uses in the Construction Industry

The multibillion dollar construction industry encompasses areas such as commercial buildings, residential homes, highways, bridges, and water and sewer systems. These areas, which we often take for granted, would not be possible without ceramic materials. Products such as floor, wall and roofing tile, cement, brick, gypsum, sewer pipe, and glass are the building blocks in the world of construction.

What would our homes and businesses be if we did not have windows? The applications of glass in the construction industry include various types of windows to let in natural light. Approximately three billion square feet of glass is produced each year to make various types of windows. To put things in perspective, imagine a 200-foot wide glass highway stretching from New York to Los Angeles. The different types of glass for windows include safety, stained, tinted, laminated and non-reflective. Additionally, glass fibers are used for insulation, ceiling panels and roofing shingles, helping us stay warm and dry.

Clay brick is used to build homes and commercial buildings because of its strength, durability and beauty. Brick is the only building product that will not burn, melt, dent, peel, warp, rot, rust or be eaten by termites. Brick comes in approximately 10 000 different colors, textures, sizes and shapes. Ceramic tile is used in applications such as flooring, walls, countertops and fireplaces. Tile also is a very durable and hygienic construction product that adds beauty to any application.

17.2.3 Lighting Electrical Appliances

An important invention that changed the lives of millions of people was the incandescent light bulb. This important invention by Thomas Edison in 1879 would not be possible without the use of glass. Glass properties of hardness, transparency, and its ability to withstand high temperatures and hold a vacuum at the same time made the light bulb a reality.

The evolution of lighting technology since this time has been characterized by the invention of increasingly brighter and more efficient light sources. By the middle of twentieth century, methods of lighting seemed well established with filament and fluorescent lamps for interiors, neon lamps for exterior advertising and signs, and sodium discharge lamps for streets. Since this time, light-emitting diode[①] (LED) technology has been developed with applications in watches, instrument panel indicators, telecommunications (optical fiber networks), data storage (CD technology) and document production (laser printers).

17.2.4 Electrical Applications

The vast electronic industry would not exist without ceramics. Ceramics have a wide range of electrical properties including insulating, semiconducting, superconducting, piezoelectric and magnetic. Ceramics are critical to products such as cell phones, computers, television and other consumer electronic products.

Magnetic storage of data has developed in parallel with semiconductor computer chips and has been equally vital to computing and information handling. Without magnetic storage there would be no Internet, no personal computers, no large databases and no gigabytes, terabytes and exabytes of data which computers now manipulate. Today more than 150 million hard disc drives and 50 million video cassette recorders are made annually worldwide.

Ceramics are even being used to enhance our sporting activities. Piezoelectric ceramics (piezoceramics) are being used to make "smart" sporting goods equipment. That is, sporting goods that can respond to its surrounding environment in order to increase its effectiveness. Uses include snow skis, baseball/softball bats and shock absorbers in mountain bicycles.

For example, K2 Corporation uses a control module produced by active control experts inside a line of its skis. The control module contains a piezoceramic material that dampens vibrations from the ice and snow, helping keep the skis on the snow

Chapter 17 Application of Materials

and thus enhancing stability, control and ultimately speed. Worth, Inc. now has a line of bats that incorporates a similar module. The piezoceramic control module helps to increase the bats sweet spot and reduces unwanted sting on off-center hits.

Ceramic spark plugs, which are electrical insulators, have had a large impact on society. They were first invented in 1860 to ignite fuel for internal combustion engines and are still being used for this purpose today. Applications include automobiles, boat engines, lawnmowers and the like. High voltage insulators make it possible to safely carry electricity to houses and businesses.

17.2.5 Communication

Glass optic fibers have provided a technological breakthrough in the area of telecommunications. Information that was once carried electrically through hundreds of copper wires can now be carried through one high-quality, transparent, silica (glass) fiber. Using this technology has increased the speed and volume of information that can be carried by orders of magnitude over that which is possible using copper cable.

Optical fibers are a reliable conduit for delivering an array of interactive services, using combinations of voice, data and video. Whether it's multimedia and video applications, high-speed data transmissions and Internet access, telecommuting or sophisticated, on-demand services, optical fibers make it easier to communicate.

17.2.6 Medical

Ceramics are becoming increasingly useful to the medical world. Surgeons are using bioceramic materials for repair and replacement of human hips, knees and other body parts. Ceramics also are being used to replace diseased heart valves. When used in the human body as implants or even as coatings to metal replacements, ceramic materials can stimulate bone growth, promote tissue formation and provide protection from the immune system.

Dentists are using ceramics for tooth replacement impants and braces. Glass microspheres smaller than a human hair are being used to deliver large, localized amounts of radiation to diseased organs in the body. Ceramics are one of the few materials that are durable and stable enough to withstand the corrosive effect of bodily fluids.

Imaging systems are critical for medical diagnostics. Modern ceramic materials play an important role in both ultrasonic and X-ray computed tomography (CT) systems. Transducers utilizing; lead zirconate titanate (PZT) based piezoelectric ceramics are the heart of ultrasonic systems. Transducers utilizing generate the ultrasonic acoustic waves and detect the reflected signals to form the image.

Ultrasound can be used to examine many parts of the body including the abdomen, breasts, female pelvis, prostate, thyroid and parathyroid, and the vascular system. Most commonly, ultrasound is used during the first, second, or third trimester of pregnancy. New developments in ultrasound now enable doctors to see reliable images of ulcerated plaque or irregularities in the blood vessels-something doctors have never been able to see before. This may allow doctors to better identify those at risk for stroke because of its ability to more clearly illustrate the flow of blood, the walls of the carotid artery and the movement of plaque.

X-ray CT scans are now a common diagnostic procedure in hospitals and clinical sites to image selected regions inside the human body for detection of cancer and other diseases. For CT scans, an X-ray detector is a crucial component that must have high efficiency to obtain high quality images. In 1988 GE Medical Systems unveiled an ultrahigh-performance detector with a breakthrough ceramic scintillator which gives better images using lower X-ray doses to the patient.

17.2.7 Environmental and Space Applications

Ceramics play an important role in addressing various environmental needs. Ceramics help decrease pollution, capture toxic materials and encapsulate nuclear waste. Today's catalytic converters in trucks and cars are made of cellular ceramics and help convert noxious hydrocarbons and carbon monoxide gases into harmless carbon dioxide and water. Advanced ceramic components are starting to be used in diesel and automotive engines. Ceramics light-weight, high-temperature and wear resistant properties results in more efficient combustion and significant fuel savings. Ceramics also are used in oil spill containment booms that corral oil so it can be towed away from ships, harbors or offshore oil drilling rigs before being burned off safely.

Reusable, lightweight and ceramic tile make NASA's space shuttle program possible. The 34 000 thermal barrier tiles protect the astronauts and the shuttles aluminum frame from the extreme temperatures (up to approximately 1 600 ℃) encountered upon reentry into the earth's atmosphere.

While the list could go on, one can begin to grasp the critical impact that ceramics have played in our past and present and will continue to play in the future.

Chapter 17　Application of Materials

17.3　Application of Polymeric Materials

17.3.1　Power of Plastic

Despite many desirable properties of plastics, plastics have always been dependent on metals as a source of power. Mobile phones, laptop computers, hand-held video games all have needed metal compounds to generate the electricity they need to work. Metals have many problems associated with them. Many of them, such as the lithium used in watch batteries, are very dangerous to work with. The manufacture of the batteries is not a task to be treated lightly. Most of them, such as cadmium in the Ni-cad rechargeable batteries, are also highly toxic, and pose a serious problem in terms of environmentally friendly disposal. Of course, all these problems lead to an inevitable increase in the cost of manufacture, which must be met by the consumer.

Now it seems that a new alternative to these costly heavy metal systems is on the horizon. Imagine instead a battery which could be made from totally recyclable plastics. A battery which incorporates only nontoxic materials, and so poses no danger to people. A battery which could be moulded to any desired shape or design; one which could be used as part of the body of a mobile phone or a computer.

The technology for such a battery is being worked on by Peter Searson and Ted Poeheler at the John Hopkins campus in Maryland, Balitmore. Preliminary tests seem to show that the plastic batteries can outperform their metallic conmterparts in terms of how much energy they can store, the voltages they can produce, the temperatures they can operate at, and how often they can be recharged.

Batteries or electrochemical cells have been around for many years, and the principles of their operation have remained the same since their discovery. They consist of three parts: the electrolyte through which charged ions move inside the cell to allow the chemical reaction to produce electricity, and two electrodes. One of these, the anode, is broken down during the reaction to produce the ions and electrons. The electrons travel out of the cell and into the wires of the system that is using the electricity. The other electrode, the cathode, takes in electrons from the system and recombines them with the ions released from the anode, thus completing the circuit. So far, the electrodes have nearly always been built from metals, since metals can easily and reversibly be broken down into the electrons and ions.

Chapter 17 Application of Materials

For the past 25 years or so, chemists have known about plastics which can conduct electricity. They consist of long polymer chains, which have a long chain of interacting electrons running along their length. This itself dose not conduct; it is like a rode packed with bumper cars to bumper nobody can go anywhere. What can be done is to remove an electron from the chain, and so creat a hole. Like removing one car from the traffic jam, this enables the electrons to move about, allowing electricity to flow.

In principle, therefore, it became possible to make electrode out of plastics; by removing an electron from the chain, the plastic became conducting and able to receive an electron from outside, in the same way as a normal cathode. This is called p-type doping, because the loss of the electron from the chain gives it a positive (p) charge.

The problem came when people tried to make the plastic equivalent of an anode; this required the addition of an electron to the chain (n-type doping), rather than the removal of one. Unfortunately, the plastics always became unstable when the electron was added. Nobody could find an n-type plastic which would work.

In the late summer of 1995, Searson and Poehler made an important discovery. They found that combining two types of plastic, polypyrrole and polystyrene sulphonate, gave a product which was stable when an electron was added. However, after doing tests on this materials, which could be used an anode in a battery, they discarded it. Two problems had arisen. The voltage produced by the new anode could not be raised above IV; and the substance which was being used to supply the extra electron to the plastic was the dangerously reactive metal lithium.

But the discovery of this plastic had been a significant step forward. By the start of January 1996, Searson and Poehler had produced a new plastic, this time based on polythiophene. Tests showed that this plastic could be used as an anode, to form a battery producing voltages much higher than the single volt of its predecessor. More importantly, the agent used to dope the plastic had been changed. No longer lithium, the team was now using a nonmetallic material.

In July 1996, the world's first all-plastic battery was produced, using the new anode. Tests showed that it did indeed exceed many of the specifications of existing metallic batteries.

Exact details of nature of the plastic, as well as the compound used to dope it with the extra electron, have not been revealed by the team since the technology has not yet been patented. In addition, the new battery is not without its problems. Calculations suggest that, although the battery is lighter, it does not store as much charge as some of the lithium based products available in using the plastic cell.

More importantly, the prototype loses charge at a rate of 2 % per week if left standing around, compared to the 0.2 % per week of the lithium battery. Though the inventors say that this will not be a problem for batteries which are in constant use, since they must be recharged regularly anyway, they are looking into the cause of the reduced storage efficiency to see what improvements can be made.

In the world of today, the implications are tremendous. Many battery manufacturers are lining up to a look at the all-plastic design. Before long, the power of plastic could be upon us.

17.3.2 Greener Plastics from a Green House Gas

Over the past decade we have all become increasingly aware of the great problems of pollution and of energy consumption, but the industries which produce these pollutants and use up our natural resources of fossil fuels are those which manufacture the day to day goods that we would now find it difficult to live without. A quick look around you will show just how much we rely on plastic, but their mass production means using and producing many chemicals that are harmful to the environment.

To see why this is the case, we need to know how plastics are made. Plastics are a type of polymer: long chain-like molecules made by linking lots of smaller molecules (called monomers) together, like coaches in a train. The chemical reactions needed to do this have to be carried out in a solvent: a liquid that dissolves things, just as sugar dissolves in tea. This solvent is either water, which ends up contaminated or petro-like compound made from fossil fuels, sometimes the CFCs (chlorofluorocarbons) which damage the ozone layer.

So polymer production delivers a double whammy, using up fossil fuels for energy and as solvents. And although recycling plastics may sound like a good idea, it actually takes more energy than making new ones. To continue our present lifestyle we clearly need a cleaner, greener way of making plastics.

Chemists all over the world are working to come up with alternatives to the conventional solvents, and in the past few years some have suggested a surprising replacement: carbon dioxide. Know to chemists as CO_2, this simple molecule forms part of the air we breathe.

But how can this gas replace the liquids normally used? Although we think of CO_2 as a gas, it can exist as a liquid or even a solid (dry ice) if we change the pressure or temperature; but even more useful is CO_2 in the form knows as a "supercritical fluid". In this state it can dissolve the chemicals needed to make polymers.

A supercritical fluid is neither liquid nor gas, but a hybrid of the two. It is formed by raising the temperature and the pressure at the same time. Raising the temperature gives the molecules the energy to move around fast, as in a gas, and so the fluid has the viscosity of a gas. But raising the pressure as well keeps the molecules close together as though they are in a liquid, and so the fluid has the density of a liquid and is able to dissolve things.

Much of the work on making polymers in CO_2 has been carried out by Joe De Simone and his colleagues at the University of North Carolina. In 1992, he reported in the magazine science that he had made a fluorine-containing polymer in supercritical CO_2. His success caused great excitement because making fluorine-containing polymers normally means using CFCs which are now banned in many countries. The added bonus is that at the end of the production, the CO_2 can simply be evaporated off as a pure gas and there is no contaminated waste solvent to be disposed of.

Unfortunately it is not as easy to make polymers that don't contain fluorine because although the monomers are soluble in CO_2 the polymers are not. This means that the growing polymer chains all group together in a big sticky mess and end up with chains of very different lengths which are useless for making plastic objects.

This is also a common problem with polymer synthesis in water or other solvents and the answer is the same one as we use to get grease off our plates and into the washing-up water: using a detergent (also known as surfactant). Fat molecules don't like to dissolve in water but the molecules in washing-up liquid have two different ends: one that likes water and one that likes grease. This breaks the grease up, enabling balls to form that are surrounded by detergent molecules, with their grease-loving ends next to the grease and the water-loving ends sticking out into the water holding the grease ball in solution.

Joe De Simone has now successfully made other polymers in CO_2 using a surfactant. Instead of having two ends like washing-up liquid, is surfactant has a backbone that likes to be next to the new polymer, with spines sticking out of it that likes being in CO_2 and enable the growing polymer to form balls suspended in the CO_2. Different concentrations of surfactant enable different sized polymers balls, and therefore different length chains and particular kinds of plastic to be made.

The clever thing about this method is that Joe De Simone's surfactant is actually his first polymer, the fluorine-containing polymer, so supercritical CO_2 is used in the making of both the polymer and the surfactant. It can also be used to separate the polymers from and left over monomer or other contaminants, making the use of CFCs and petrol-like solvents completely unnecessary.

But wait! Isn't CO_2 that nasty gas that causes global warming? Surely we don't want any more of that around? Well thankfully we don't end to, the reverse is true. The CO_2 used is taken straight from the atmosphere and used over and over again and so is one of the cheapest and most readily available chemicals known. Thanks to chemists, the gas we normally think of as a baddie can be turned into a goodies to help clean up the world we live in.

17.4 Application of Composite Materials

17.4.1 Defense and Military Industry

Composite materials have widely applied in the areas of defense and military industry, due to its typical lightweight property, high specific strength and modulus, material properties of designability, and other excellent properties. The progress of the composites has laid a solid foundation for weapon system selection and product design, and its application has made a significant contribution to the weapon system which is realizing the lightweight, fast reaction ability, high strength, large range, precision strikes, *etc*.

In some countries, the missiles and rockets have achieved the structure of composite materials. They become lightweight and miniaturization, and advance to new and higher level. Such as the former Soviet Union's anti-tank missile (AT-3), its main composite structures are: cap, shell, tail seat, tail, *etc*. These composite structures take the accounts for 75 % (Figure 17.1).

Figure 17.1 Anti-tank missile

The United States Allied-Signals Company successfully developed a high-performance bulletproof nonwovens composite material (spectra shield). With its lightweight, great flexibility, good protective effect and other excellent performance, it has been applied in the field of bulletproof products, such as body armor (Figure 17.2).

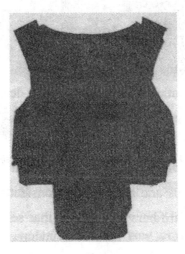

Figure 17.2 Body armor

In China, the composite material technology and the level of application technology have developed quickly, more and more composite materials are applied to weapons. Such as aramid fiber reinforced composite material has been used as the armor of main battle tank.

17.4.2 Aerospace Vehicle

The potential benefits of emerging technologies to aeronautics and their foundation in composite materials are described and the resulting benefits in vehicle take off gross weight are quantified. Finally, a 21st Century vision for aeronautics in which human mobility is increased by an order of magnitude is articulated. Advanced composites have emerged as the structural materials of choice for many aerospace applications because of their superior specific strength and stiffness properties. First developed for military aircraft applications, composites now play a significant role in a broad range of current generation military aerospace systems. Commercial transport aviation has also witnessed a significant increase in adoption of composites during the past ten year.

Composite material has become the main structure material of satellite (Figure 17.3). The application of composite materials reduces the satellites' total quality and the

Chapter 17 Application of Materials

product cost significantly and increases the payload. Many structures are made of composites, such as antenna, solar cell, center bearing cylinder, and so on.

Figure 17.3 Composite materials used in satellite

The airbus A380 uses lots of light composite materials, such as the reduction plate, vertical and horizontal stabilizer (used as fuel tank), rudder, elevator, landing gear doors, fairing, vertical tail box, rudder, elevator, upper cabin floor beam, horizontal tail and aileron, *etc*. The use of composites takes about 25 % among all materials (Figure 17.4).

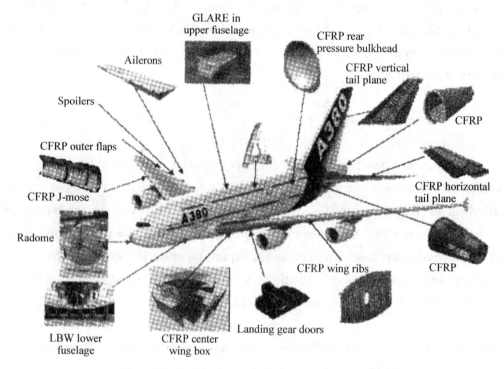

Figure 17.4 Composite materials used in airbus A380

17.4.3 Transport

As composite material is auto's mainstream lightweight material, its application has been rapidly developing in transportation field. No matter in developed countries or fast developing countries, composite materials have a large number of applications in automobile production. The main application range is exterior trim, structural components, half structure, *etc*, (Figure 17.5). Physical structures are the car door, window, body, insulation board, bumper, *etc*. In recent years, attempts have been observed to reduce the use of expensive glass, aramid or carbon fibers and also lighten considerably the cars body by taking advantage of the lower density and cost that some natural fibers provide. In that sense, renewable fibers as reinforcements were vastly used in composites of interior parts for a number of passenger and commercial vehicles. The natural fiber reinforced composites have been used as auto interior material to decrease noise.

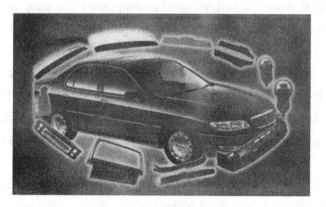

Figure 17.5 FRP used in car

In vessel transportation, composite materials are the important material for ship. At present the fiber reinforced resin matrix composites is the main material. Since it has the traditional building materials' incomparable properties, such as light high strength, good entirety and simple forming. However, due to the low elastic modulus and molding technology restricted, the ship size is restricted. The current main products are all kinds of small civilian and military ships, such as the craft fishing boat, gave birth to boat and reverse mine ship, *etc*. Figure 17.6 show ayacht made of FRP.

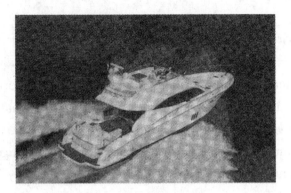

Figure 17.6 A yacht made of FRP

17.4.4 Energy Saving and Environment Protect Field

In the energy industry field, resin matrix composites are widely used in the world. Its main applications are as following.

In the thermal power industry aspect, composites are used in the ventilation system, row coal ash pipes, circulating water cooling system, roof axial flow fan, cable protection facilities, electric insulation products, *etc*.

In hydroelectric power industry aspect, many structures are made of composites, such as power station construction, dam and tunnels impingement, wear-resisting, antifreeze, corrosion resistance through the water protection; valve; power generation and transmission of all kinds of electric insulation products, *etc*.

In new energy aspect, resin matrix composites have been applied to wind generator blade, pole and electric insulation products, *etc* (Figure 17.7). China has become the largest wind energy market, the total installed capacity by the end of 2010 has reached 44 733 MW, and the added capacity in the year of 2010 is up to

Figure 17.7 FRP made blade used in wind generatior

18 900 MW which is account for 50 % of global addition; It is expected that the total installed capacity in 2020 would be 200 000 MW.

In the environment protect fields, green composites deriving from renewable resources bring very promising potential to provide benefits to companies, natural environment and end-customers due to dwindling petroleum resources. Composites made of renewable materials have been rampantly used in interior and exterior body parts. The application of green composites in automobile body panels seems to be feasible as far as green composites have comparable mechanical performance with the synthetic ones. Similar components are used as trim parts: in dashboards, door panels, parcel shelves, seat cushions, backrests and cabin linings. In recent years there has been increasing interest in the replacement of fiber glass in reinforced plastic composites by natural plant fibers such as jute, flax, hemp, sisal and ramie. China has strongly accepted advanced fabrication technology and equipment to promote the production efficiency and quality, reduce materials and energy consumption, reduce the impact to the environment, and industrialize the recycle technology for composite materials.

Notes

① light-emitting diode 发光二极管

Vocabulary

abrasion [əˈbreɪʒ(ə)n] n. 磨损;磨耗;擦伤

aeronautics [eərəˈnɔːtɪks] n. 航空学;飞行术

airbus [ˈe(ə)rˌbəs] n. 空中巴士(指大型中短程喷气客机)

artery [ˈɑːtəri] n. 动脉;干道;主流

astronaut [ˈæstrənɔːt] n. 宇航员,航天员;太空旅行者

bulletproof [ˈbʊlɪtpruːf] adj. 防弹的

carotid [kəˈrɒtɪd] n. 颈动脉 adj. 颈动脉的

ceiling [ˈsiːlɪŋ] n. 天花板;上限

clutch [klʌtʃ] n. 离合器;控制;手;紧急关头 vt. 抓住;紧握 adj. 没有手提带或背带的;紧要关头的 vi. 攫;企图抓住

countertop [ˈkaʊntətɒp] n. 工作台面

cushion [ˈkʊʃ(ə)n] n. 垫子;起缓解作用之物;(猪等的)臀肉;银行储蓄 vt. 给……安上垫子;把……安置在垫子上;缓和……的冲击

dent [dent] n. 凹痕;削弱;减少;齿 vt. 削弱;使产生凹痕 vi. 产生凹陷;凹进去;削减

dental [ˈdent(ə)l] n. 齿音 adj. 牙科的;牙齿的,牙的

designability [dezɪgneɪˈbɪlɪti] n. 设计性,结构性;可设计性

diagnostic [daɪəgˈnɒstɪk] n. 诊断法;诊断结论 adj. 诊断的;特征的

Chapter 17 Application of Materials

diesel ['diːz(ə)l] n. 柴油机；柴油；(俚)健康的身体 adj. 内燃机传动的；供内燃机用的

diode ['daɪəʊd] n. [电子]二极管

dwindling ['dwɪndlɪŋ] adj. 逐渐减少的；v. 逐渐变少或变小(dwindle 的现在分词)

encapsulate [ɪn'kæpsjʊleɪt] vt. 压缩；将……装入胶囊；将……封进内部；概述 vi. 形成胶囊

gigabyte ['gɪgəbaɪt] n. 十亿字节；十亿位组

hemp [hemp] n. 大麻；麻类植物；大麻烟卷 adj. 大麻类植物的

hygienic [haɪ'dʒiːnɪk] adj. 卫生的，保健的；卫生学的

ignition [ɪg'nɪʃ(ə)n] n. 点火，点燃；着火，燃烧；点火开关，点火装置

jute [dʒuːt] n. 黄麻；黄麻纤维

laptop ['læptɒp] n. 笔记型电脑，笔记本电脑，便携式电脑

lawnmower ['lɔːnməʊə] n. [建]剪草机

myriad ['mɪrɪəd] n. 无数，极大数量；无数的人或物 adj. 无数的；种种的

nonwoven [nɒn'wəʊvən] n. 非纺织而成的纺织品 adj. 非纺织的

noxious ['nɒkʃəs] adj. 有害的；有毒的；败坏道德的；讨厌的

ozone ['əʊzəʊn] n. [化学]臭氧；新鲜空气

plaque [plæk] n. 匾；血小板；饰板

pregnancy ['pregnənsɪ] n. 怀孕；丰富，多产；意义深长

ramie ['ræmɪ] n. 苎麻；苎麻纤维

rechargeable [riː'tʃɑːdʒəbl] adj. 可再充电的；收费的

rot [rɒt] n. 腐烂；腐败；腐坏 vt. 使腐烂；使腐朽；使堕落 vi. 腐烂；腐败；堕落

scintillator ['sɪntɪleɪtə] n. 发出闪光的东西；[核]闪烁计数器

sewer ['suːə] n. 下水道；阴沟；裁缝师 vt. 为……铺设污水管道；用下水道排除……的污水 vi. 清洗污水管

shuttle ['ʃʌt(ə)l] n. 航天飞机；穿梭；梭子；穿梭班机、公共汽车等 vt. 使穿梭般来回移动；短程穿梭般运送 vi. 穿梭往返

sisal ['saɪs(ə)l] n. 剑麻；西沙尔麻

subsonic [sʌb'sɒnɪk] n. 亚音速飞机 adj. 次音速的；比音速稍慢的

surgeon ['sɜːdʒ(ə)n] n. 外科医生

terabyte ['terəbaɪt] n. 太字节；兆兆位(量度信息单位)

termite ['tɜːmaɪt] n. [昆]白蚁

tomography [tə'mɒgrəfɪ] n. X 线断层摄影术(等于 laminography)

treatable ['triːtəbl] adj. 能治疗的；好对付的；能处理的

trimester [traɪ'mestə] n. 三个月；一学期

ventilation [ˌventɪ'leɪʃ(ə)n] n. 通风设备；空气流通

warp [wɔːp] n. 弯曲，歪曲；偏见；乖戾 vt. 使变形；使有偏见；曲解 vi. 变歪，变弯；曲解

weldable ['weldəbl] adj. [机]可焊的

whammy ['wæmɪ] n. 晦气；剧烈的打击

Exercises

1. Translate the following Chinese phrases into English

(1) 工具钢 (2) 高纯陶瓷 (3) 贵金属
(4) 排污系统 (5) 可回收塑料 (6) 臭氧层
(7) 可再生资源

2. Translate the following English phrases into Chinese

(1) high rigidity (2) high catalytic activity
(3) refractory material (4) glass fibers
(5) piezoceramic materials (6) energy consumption
(7) environment protect (8) green composites
(9) advanced fabrication technology

3. Translate the following Chinese sentences into English

(1) 如果铁轨加工时未采取相应处理，也会发生这样的弯曲。

(2) 这导致谁先谁后的状况，即一种新材料不能应用因为它价格太昂贵，它的价格很高主要因为没有大批量生产。

(3) 早期研究已表明磁性含碳纳米复合材料在磁性数据存储领域显示了巨大的应用前景。

(4) 化学产品的种类非常广，这些产品对于我们日常的生活有着无法估计的贡献，制造这些化学产品设备的生产能力小到每年几吨，大到石化产品的每年50万吨。

(5) 粒子没有择优取向性，其形状不如纤维重要。

(6) 工程师或科学家对材料的各种性质、结构与功能之间的关系以及生产工艺越熟悉，就越能熟练自信地根据这些标准选择出最合适的材料。

4. Translate the following English sentences into Chinese

(1) Composite materials in other highly visible applications include racing car bodies which have been found to provide more safety to the drivers, and longer-lasting rigidity than an older material like aluminum.

(2) The richness of electronic properties of carbon-based nanostructures including carbon nanotubes and graphene attract great interest as they have high mobility, electric conductivity robustness and environmental stability.

(3) With these challenges however, there are an equal number of opportunities to discover and apply new chemistry, to improve the economics of chemical manufacturing and to reduce the much-tarnished irnage of chemistry.

(4) The smaller the size of filler particles is, the larger their specific surface area becomes, and the more likely the agglomeration of the particles. This may lead to properties of nanoparticles filled composites even worse than conventional particle polymer systems.

(5) At this point, materials utilization was totally a selection process that involved deciding from a given, rather limited set of materials the one best suited for an application by virtue of its characteristics.

5. Translate the following Chinese essay into English

这要求必须有新的技术来降低原料、化学产品生产过程及产品本身对环境的污染；降低和减少有害化学物质的排放；最大程度地利用可循环原料，提供使用寿命长、可循环产品。

6. Translate the following English essay into Chinese

Nanotechnology encompasses a broad range of tools, techniques and applications. This technology attempts to manipulate materials at the nanoscale in order to yield novel properties that do not exist at larger scales. These novel properties may enable new or improved solutions to problems that have been challenging to solve with conventional technology. For developing countries, these solutions may include more effective and inexpensive water purification devices, energy sources, medical diagnostic tests and drug delivery systems, durable building materials, and other products. It is quite apparent that there are in numerable potential benefits for society, the environment and the world.

扫一扫，查看更多资料

Main References

[1] 杨含离,杨鄂川,刘妤.工程流体力学双语教程[M].北京:国防工业出版社,2015.

[2] 王志军,袁东升,宋文婷.工程热力学与传热学(双语)[M].徐州:中国矿业大学出版社,2015.

[3] Holman J. P. Heat Transfer (10th edition.) [M]. Reprint. Originally published: New York: McGraw-Hill, 2010.

[4] 余万华.金属材料成形双语教程(英文)[M].北京:冶金工业出版社,2012.

[5] Joel R. Fried. Polymer Science and Technology (2nd edition)[M]. Prentice-Hall International, Inc., 2003.

[6] 刘瑛,阎昱.材料成型及控制工程专业英语[M].北京:机械工业出版社,2015.

[7] 刘爱国.材料科学与工程专业英语[M].哈尔滨:哈尔滨工业大学出版社,2015.

[8] 胡礼木,王卫卫.材料成型及控制工程专业英语阅读[M].北京:机械工业出版社,2014.

[9] 刘东平.材料物理(双语)[M].大连:大连理工大学出版社,2014.

[10] 水中和.材料概论(双语)(第二版)[M].武汉:武汉理工大学出版社,2012.

[11] 刘科高,田清波.材料科学与工程专业英语精读[M].冶金工业出版社,2012.

[12] 陈克正,王玮,刘春廷.材料科学与工程导论(双语)[M].北京:化学工业出版社,2011.

[13] William F. Smith, Javad Hashemi. Foundations of Materials Sciemce and Engineering [M]. Prentice-Hall International, Inc., 2010.

[14] 郝凌云,陈晓玉,叶原丰.复合材料与工程专业英语[M].北京:化学工业出版社,2014.

[15] 材料科学与工程抓野英语编委会.材料科学与工程专业英语[M].北京:中国石化出版社,2011.

[16] William D. Callister, Jr. Fundamentals of Materials Science and Engineering (5th edition)[M]. 北京:化学工业出版社,2006.

[17] 黄培彦.材料科学与工程导论(双语)[M].广州:华南理工大学出版社,2007.

[18] 赵杰.材料科学基础(双语版)[M].大连:大连理工大学出版社,2015.

[19] 张耀君.纳米材料基础(双语版)[M].北京:化学工业出版社,2013.

[20] 匡少平,王世颖.材料科学与工程专业英语[M].北京:化学工业出版社,2013.

Main References

[21] 范小红,徐勇.材料成型及控制工程专业英语教程[M].北京:化学工业出版社,2014.

[22] 王快社,刘环,张郑.材料加工工程科技英语[M].北京:冶金工业出版社,2013.